Advanced Mechanics

Advanced Mechanics

From Euler's Determinism to Arnold's Chaos

S. G. Rajeev

Department of Physics and Astronomy, Department of Mathematics
University of Rochester, Rochester, NY14627

OXFORD
UNIVERSITY PRESS

Great Clarendon Street, Oxford, OX2 6DP,
United Kingdom

Oxford University Press is a department of the University of Oxford.
It furthers the University's objective of excellence in research, scholarship,
and education by publishing worldwide. Oxford is a registered trade mark of
Oxford University Press in the UK and in certain other countries

The moral rights of the author have been asserted

First Edition published in 2013

Impression: 1

Published in the United States of America by Oxford University Press
198 Madison Avenue, New York, NY 10016, United States of America

British Library Cataloguing in Publication Data

Data available

Library of Congress Control Number: 2013938985

ISBN 978-0-19-967085-7
ISBN 978-0-19-967086-4 (pbk.)

Printed and bound by
CPI Group (UK) Ltd, Croydon, CR0 4YY

This book is dedicated to Sarada Amma and Gangadharan Pillai, my parents.

Preface

Classical mechanics is the oldest and best understood part of physics. This does not mean that it is cast in marble yet, a museum piece to be admired reverently from a distance. Instead, mechanics continues to be an active area of research by physicists and mathematicians. Every few years, we need to re-evaluate the purpose of learning mechanics and look at old material in the light of modern developments.

The modern theories of chaos have changed the way we think of mechanics in the last few decades. Previously formidable problems (three body problem) have become easy to solve numerically on personal computers. Also, the success of quantum mechanics and relativity gives new insights into the older theory. Examples that used to be just curiosities (Euler's solution of the two center problem) become starting points for physically interesting approximation methods. Previously abstract ideas (Julia sets) can be made accessible to everyone using computer graphics. So, there is a need to change the way classical mechanics is taught to advanced undergraduates and beginning graduate students.

Once you have learned basic mechanics (Newton's laws, the solution of the Kepler problem) and quantum mechanics (the Schrödinger equation, hydrogen atom) it is time to go back and relearn classical mechanics in greater depth. It is the intent of this book to take you through the ancient (the original meaning of "classical") parts of the subject quickly: the ideas started by Euler and ending roughly with Poincaré. Then we take up the developments of twentieth century physics that have largely to do with chaos and discrete time evolution (the basis of numerical solutions).

Although some knowledge of Riemannian geometry would be helpful, what is needed is developed here. We will try to use the minimum amount of mathematics to get as deep into physics as possible. Computer software such as Mathematica, Sage or Maple are very useful to work out examples, although this book is not about computational methods.

Along the way you will learn about: elliptic functions and their connection to the arithmetic-geometric-mean; Einstein's calculation of the perihelion shift of Mercury; that spin is really a classical phenomenon; how Hamilton came very close to guessing wave mechanics when he developed a unified theory of optics and mechanics; that Riemannian geometry is useful to understand the impossibility of long range weather prediction; why the *maximum* of the potential is a stable point of equilibrium in certain situations; the similarity of the orbits of particles in atomic traps and of the Trojan asteroids.

By the end you should be ready to absorb modern research in mechanics, as well as ready to learn modern physics in depth.

The more difficult sections and problems that you can skip on a first reading are marked with asterisks. The more stars, the harder the material. I have even included some problems whose answers I do not know, as research projects.

Mechanics is still evolving. In the coming years we will see even more complex problems solved numerically. New ideas such as renormalization will lead to deeper theories of chaos. Symbolic computation will become more powerful and change our very definition of what constitutes an analytic solution of a mechanical problem. Non-commutative geometry could become as central to quantum mechanics as Riemannian geometry is to classical mechanics. The distinction between a book and a computer will disappear, allowing us to combine text with simulations. A mid-twenty-first century course on mechanics will have many of the ingredients in this book, but the emphasis will be different.

Acknowledgements

My teacher, A. P. Balachandran, as well as my own students, formed my view of mechanics. This work is made possible by the continued encouragement and tolerance of my wife. I also thank my departmental colleagues for allowing me to pursue various directions of research that must appear esoteric to them.

For their advice and help I thank Sonke Adlung and Jessica White at Oxford University Press, Gandhimathi Ganesan at Integra, and the copyeditor Paul Beverley. It really does take a village to create a book.

Contents

List of Figures

1
The variational principle

Many problems in physics involve finding the minima (more generally extrema) of functions. For example, the equilibrium positions of a static system are the extrema of its potential energy; stable equilibria correspond to local minima. It is a surprise that even dynamical systems, whose positions depend on time, can be understood in terms of extremizing a quantity that depends on the paths: the action. In fact, all the fundamental physical laws of classical physics follow from such variational principles. There is even a generalization to quantum mechanics, based on averaging over paths where the paths of extremal action make the largest contribution. In essence, the calculus of variations is the differential calculus of functions that depend on an infinite number of variables. For example, suppose we want to find the shortest curve connecting two different points on the plane. Such a curve can be thought of as a function $(x(t), y(t))$ of some parameter (like time). It must satisfy the boundary conditions

$$x(t_1) = x_1, y(t_1) = y_1$$
$$x(t_2) = x_2, y(t_2) = y_2$$

where the initial and final points are given. The length is

$$S[x, y] = \int_{t_1}^{t_2} \sqrt{\dot{x}^2 + \dot{y}^2} dt$$

This is a function of an infinite number of points because we can make some small changes $\delta x(t)$, $\delta y(t)$ at each time t independently. We can define a differential, the infinitesimal change of the length under such a change:

$$\delta S = \int_{t_1}^{t_2} \frac{\dot{x}\delta\dot{x} + \dot{y}\delta\dot{y}}{\sqrt{\dot{x}^2 + \dot{y}^2}} dt$$

Generalizing the idea from the calculus of several variables, we expect that at the extremum, this quantity will vanish for any $\delta x, \delta y$. This condition leads to a differential equation whose solution turns out to be (no surprise) a straight line. There are two key ideas here. First of all, the variation of the time derivative is the time derivative of the variation:

$$\delta\dot{x} = \frac{d}{dt}\delta x$$

This is essentially a postulate on the nature of the variation. (It can be further justified if you want.) The second idea is an integration by parts, remembering that the variation must vanish at the boundary (we are not changing the initial and final points).

$$\delta x(t_1) = \delta x(t_2) = 0 = \delta y(t_1) = \delta y(t_2)$$

Now,

$$\frac{\dot{x}}{\sqrt{\dot{x}^2 + \dot{y}^2}} \frac{d}{dt} \delta x = \frac{d}{dt} \left[\frac{\dot{x}}{\sqrt{\dot{x}^2 + \dot{y}^2}} \delta x \right] - \frac{d}{dt} \left[\frac{\dot{x}}{\sqrt{\dot{x}^2 + \dot{y}^2}} \right] \delta x$$

and similarly with δy. Then

$$\delta S = \int_{t_1}^{t_2} \frac{d}{dt} \left[\frac{\dot{x}}{\sqrt{\dot{x}^2 + \dot{y}^2}} \delta x + \frac{\dot{y}}{\sqrt{\dot{x}^2 + \dot{y}^2}} \delta y \right] dt$$

$$- \int_{t_1}^{t_2} \left\{ \frac{d}{dt} \left[\frac{\dot{x}}{\sqrt{\dot{x}^2 + \dot{y}^2}} \right] \delta x + \frac{d}{dt} \left[\frac{\dot{y}}{\sqrt{\dot{x}^2 + \dot{y}^2}} \right] \delta y \right\} dt$$

The first term is a total derivative and becomes

$$\left[\frac{\dot{x}}{\sqrt{\dot{x}^2 + \dot{y}^2}} \delta x + \frac{\dot{y}}{\sqrt{\dot{x}^2 + \dot{y}^2}} \delta y \right]_{t_1}^{t_2} = 0$$

because δx and δy both vanish at the boundary. Thus

$$\delta S = - \int_{t_1}^{t_2} \left\{ \frac{d}{dt} \left[\frac{\dot{x}}{\sqrt{\dot{x}^2 + \dot{y}^2}} \right] \delta x + \frac{d}{dt} \left[\frac{\dot{y}}{\sqrt{\dot{x}^2 + \dot{y}^2}} \right] \delta y \right\} dt$$

In order for this to vanish for any variation, we must have

$$\frac{d}{dt} \left[\frac{\dot{x}}{\sqrt{\dot{x}^2 + \dot{y}^2}} \right] = 0 = \frac{d}{dt} \left[\frac{\dot{y}}{\sqrt{\dot{x}^2 + \dot{y}^2}} \right]$$

That is because we can choose a variation that is only non-zero in some tiny (as small you want) neighborhood of a particular value of t. Then the quantity multiplying it must vanish, independently at each value of t. These differential equations simply say that the vector $(\dot{x}, \dot{y})$ has constant direction: $\left(\frac{\dot{x}}{\sqrt{\dot{x}^2 + \dot{y}^2}}, \frac{\dot{y}}{\sqrt{\dot{x}^2 + \dot{y}^2}} \right)$ is just the unit vector along the tangent. So the solution is a straight line. Why did we do all this work to prove an intuitively obvious fact? Because sometimes intuitively obvious facts are wrong. Also, this method generalizes to situations where the answer is not at all obvious: what is the curve of shortest length between two points that lie entirely on the surface of a sphere?

1.1.   Euler–Lagrange equations

In many problems, we will have to find the extremum of a quantity

$$S[q] = \int_{t_1}^{t_2} L[q, \dot{q}, t] dt$$

where $q^i(t)$ are a set of functions of some parameter t. We will call them position and time respectively, although the actual physical meaning may be something else in a particular case. The quantity $S[q]$, whose extremum we want to find, is called the action. It depends on an infinite number of independent variables, the values of q at various times t. It is the integral of a function (called the Lagrangian) of position and velocity at a given time, integrated on some interval. It can also depend explicitly on time; if it does not, there are some special tricks we can use to simplify the solution of the problem.

As before, we note that at an extremum S must be unchanged under small variations of q. Also we assume the identity

$$\delta \dot{q}^i = \frac{d}{dt} \delta q^i$$

We can now see that

$$\delta S = \int_{t_1}^{t_2} \sum_i \left[\delta \dot{q}^i \frac{\partial L}{\partial \dot{q}^i} + \delta q^i \frac{\partial L}{\partial q^i} \right] dt$$

$$= \int_{t_1}^{t_2} \sum_i \left[\frac{d \delta q^i}{dt} \frac{\partial L}{\partial \dot{q}^i} + \delta q^i \frac{\partial L}{\partial q^i} \right] dt$$

We then do an integration by parts:

$$= \int_{t_1}^{t_2} \sum_i \frac{d}{dt} \left[\delta q^i \frac{\partial L}{\partial \dot{q}^i} \right] dt$$

$$+ \int_{t_1}^{t_2} \sum_i \left[-\frac{d}{dt} \frac{\partial L}{\partial \dot{q}^i} + \frac{\partial L}{\partial q^i} \right] \delta q^i dt$$

Again in physical applications, the boundary values of q at times t_1 and t_2 are given. So

$$\delta q^i(t_1) = 0 = \delta q^i(t_2)$$

Thus

$$\int_{t_1}^{t_2} \sum_i \frac{d}{dt} \left[\delta q^i \frac{\partial L}{\partial \dot{q}^i} \right] dt = \left[\delta q^i \frac{\partial L}{\partial \dot{q}^i} \right]_{t_1}^{t_2} = 0$$

and at an extremum,

$$\int_{t_1}^{t_2} \sum_i \left[-\frac{d}{dt}\frac{\partial L}{\partial \dot{q}^i} + \frac{\partial L}{\partial q^i} \right] \delta q^i dt = 0$$

Since these have to be true for all variations, we get the differential equations

$$-\frac{d}{dt}\frac{\partial L}{\partial \dot{q}^i} + \frac{\partial L}{\partial q^i} = 0$$

This ancient argument is due to Euler and Lagrange, of the pioneering generation that figured out the consequences of Newton's laws. The calculation we did earlier is a special case. As an exercise, re-derive the equations for minimizing the length of a curve using the Euler–Lagrange equations.

1.2.   The Variational principle of mechanics

Newton's equation of motion of a particle of mass m and position q moving on the line, under a potential $V(q)$, is

$$m\ddot{q} = -\frac{\partial V}{\partial q}$$

There is a quantity $L(q,\dot{q})$ such that the Euler–Lagrange equation for minimizing $S = \int L[q,\dot{q}]dt$ is just this equation.

We can write this equation as

$$\frac{d}{dt}[m\dot{q}] + \frac{\partial V}{\partial q} = 0$$

So if we had

$$m\dot{q} = \frac{\partial L}{\partial \dot{q}}, \quad \frac{\partial L}{\partial q} = -\frac{\partial V}{\partial q}$$

we would have the right equations. One choice is

$$L = \frac{1}{2}m\dot{q}^2 - V(q)$$

This quantity is called the Lagrangian. Note that it is the *difference* of kinetic and potential energies, and not the sum. More generally, the coordinate q may be replaced by a collection of numbers q^i, $i = 1, \cdots, n$ which together describe the instantaneous position of a system of particles. The number n of such variables needed is called the number of degrees of freedom. Part of the advantage of the Lagrangian formalism over the older Newtonian one is that it allows even curvilinear co-ordinates: all you have to know are the kinetic energy and potential energy in these co-ordinates. To be fair, the Newtonian formalism is more general in another direction, as it allows forces that are not conservative (a system can lose energy).

Example 1.1: The kinetic energy of a particle in spherical polar co-ordinates is

$$\frac{1}{2}m\left[\dot{r}^2 + r^2\dot{\theta}^2 + r^2\sin^2\theta\dot{\phi}^2\right]$$

Thus the Lagrangian of the Kepler problem is

$$L = \frac{1}{2}m\left[\dot{r}^2 + r^2\dot{\theta}^2 + r^2\sin^2\theta\dot{\phi}^2\right] + \frac{GMm}{r}$$

1.3. Deduction from quantum mechanics*

Classical mechanics is the approximation to quantum mechanics, valid when the action is small compared to Planck's constant $\hbar \sim 6 \times 10^{-34}\text{m}^2$ kg s^{-1}. So we should be able to deduce the variational principle of classical mechanics as the limit of some principle of quantum mechanics. Feynman's action principle of quantum mechanics says that the probability amplitude for a system to start at q_1 at time t_1 and end at q_2 at time t_2 is

$$K(q', q|t) = \int_{q(t_1)=q_1}^{q(t_2)=q_2} e^{\frac{i}{\hbar}S[q]}\mathcal{D}q$$

This is an infinite dimensional integral over all paths (functions of time) that satisfy these boundary conditions. Just as classical mechanics can be formulated in terms of the differential calculus in function spaces (variational calculus), quantum mechanics uses the integral calculus in function spaces. In the limit of small $\hbar$ the oscillations are very much more pronounced: a small change in the path will lead to a big change in the phase of the integrand, as the action is divided by $\hbar$. In most regions of the domain of integration, the integral cancels itself out: the real and imaginary parts change sign frequently. The exception is the neighborhood of an extremum, because the phase is almost constant and so the integral will not cancel out. This is why the extremum of the action dominates in the classical limit $\hbar \to 0$. The best discussion of these ideas is still in Feynman's classic paper Feynman (1948).

1.3.1   The definition of the path integral

Feynman's paper might alarm readers who are used to rigorous mathematics. Much of the work of Euler was not mathematically rigorous either: the theorems of variational calculus are from about 1930s (Sobolev, Morse et al.), about two centuries after Euler. The theory of the path integral is still in its infancy. The main step forward was by Wiener who defined integrals of the sort we are using, except that instead of being oscillatory (with the i in the exponential) they are decaying. A common trick is to evaluate the path integral for imaginary time, where theorems are available, then analytically continue to real time when the mathematicians aren't looking. Developing an integral calculus in function spaces remains a great challenge for mathematical physics of our time.

Problem 1.1: Find the solution to the Euler–Lagrange equations that minimize

$$S[q] = \frac{1}{2}\int_0^a \dot{q}^2 dt$$

subject to the boundary conditions

$$q(0) = q_0, \quad q(a) = q_1$$

Problem 1.2: Show that, although the solution to the equations of a harmonic oscillator is an extremum of the action, it need not be a minimum, even locally.

Solution
The equation of motion is

$$\ddot{q} + \omega^2 q = 0$$

which follows from the action

$$S[q] = \frac{1}{2} \int_{t_1}^{t_2} [\dot{q}^2 - \omega^2 q^2] dt$$

A small perturbation $q \to q + \delta q$ will not change the boundary conditions if δq, vanish at t_1, t_2. It will change the action by

$$S[q + \delta q] = S[q] + \int_{t_1}^{t_2} [\dot{q}\delta\dot{q} - \omega^2 q \delta q] dt + \frac{1}{2} \int_{t_1}^{t_2} [(\delta\dot{q})^2 - \omega^2(\delta q)^2] dt$$

The second term will vanish if q satisfies the equations of motion. An example of a function that vanishes at the boundary is

$$\delta q(t) = A \sin \frac{n\pi(t - t_1)}{t_2 - t_1}, \quad n \in \mathbb{Z}$$

Calculate the integral to get

$$S[q + \delta q] = S[q] + A^2 \frac{t_2 - t_1}{4} \left\{ \left(\frac{n\pi}{t_2 - t_1} \right)^2 - \omega^2 \right\}$$

If the time interval is long enough $t_2 - t_1 > \frac{n\pi}{\omega}$ such a change will lower the action. The longer the interval, the more such variations exist.

Problem 1.3: A steel cable is hung from its two end points with co-ordinates (x_1, y_1) and (x_2, y_2). Choose a Cartesian co-ordinate system with the y-axis vertical and the x-axis horizontal, so that $y(x)$ gives the shape of the chain. Assume that its weight per unit length is some constant μ. Show that the potential energy is

$$\mu \int_{x_1}^{x_2} \sqrt{1 + y'^2(x)} y(x) dx$$

Find the condition that $y(x)$ must satisfy in order that this be a minimum. The solution is a curve called a *catenary*. It also arises as the solution to some other problems. (See next chapter.)

2
Conservation laws

2.1. Generalized momenta

Recall that if q is a Cartesian co-ordinate,

$$p = \frac{\partial L}{\partial \dot{q}}$$

is the momentum in that direction. More generally, for any co-ordinate q^i the quantity

$$p_i = \frac{\partial L}{\partial \dot{q}^i}$$

is called the *generalized momentum* conjugate to q^i. For example, in spherical polar co-ordinates the momentum conjugate to ϕ is

$$p_\phi = mr^2\dot{\phi}$$

You can see that this has the physical meaning of angular momentum around the third axis.

2.2. Conservation laws

This definition of generalized momentum is motivated in part by a direct consequence of it: if L happens to be independent of a particular co-ordinate q^i (but might depend on $\dot{q}^i$), then the momentum conjugate to it is independent of time, that is, it is conserved:

$$\frac{\partial L}{\partial q^i} = 0 \implies \frac{d}{dt}\left[\frac{\partial L}{\partial \dot{q}^i}\right] = 0$$

For example, p_ϕ is a conserved quantity in the Kepler problem. This kind of information is precious in solving a mechanics problem; so the Lagrangian formalism which identifies such conserved quantities is very convenient to actually solve for the equations of a system.

2.3.   Conservation of energy

L can have a time dependence through its dependence of $q, \dot{q}$ as well as explicitly. The total time derivative is

$$\frac{dL}{dt} = \sum_i \dot{q}^i \frac{\partial L}{\partial q^i} + \sum_i \ddot{q}^i \frac{\partial L}{\partial \dot{q}^i} + \frac{\partial L}{\partial t}$$

The E-L equations imply

$$\frac{d}{dt} \left[\sum_i p_i \dot{q}^i - L \right] = -\frac{\partial L}{\partial t}, \quad p_i = \frac{\partial L}{\partial \dot{q}^i}$$

In particular, if L has no explicit time dependence, the quantity called the *hamiltonian*,

$$H = \sum_i p_i \dot{q}^i - L$$

is conserved.

$$\frac{\partial L}{\partial t} = 0 \implies \frac{dH}{dt} = 0$$

What is its physical meaning? Consider the example of a particle in a potential

$$L = \frac{1}{2} m \dot{q}^2 - V(q)$$

Since the kinetic energy T is a quadratic function of $\dot{q}$, and V is independent of $\dot{q}$,

$$p\dot{q} = \dot{q} \frac{\partial T}{\partial \dot{q}} = 2T$$

Thus

$$H = 2T - (T - V) = T + V$$

Thus the hamiltonian, in this case, is the total energy.

More generally, if the kinetic energy is quadratic in the generalized velocities $\dot{q}^i$ (which is true very often) and if the potential energy is independent of velocities (also true often), the hamiltonian is the same as energy. There are some cases where the hamiltonian and energy are not the same though: for example, when we view a system in a reference frame that is not inertial. But these are unusual situations.

2.4. Minimal surface of revolution

Although the main use of the variational calculus is in mechanics, it can also be used to solve some interesting geometric problems. A *minimal surface* is a surface whose area is unchanged under small changes of its shape. You might know that for a given volume, the sphere has minimal area. Another interesting question in geometry is to ask for a surface of minimal area which has a given curve (or a disconnected set of curves) as boundary. The first such problem was solved by Euler. What is the surface of revolution of minimal area, with given radii at the two ends? Recall that a surface of revolution is what you get by taking some curve $y(x)$ and rotating it around the x-axis. The cross-section at x is a circle of radius $y(x)$, so we assume that $y(x) > 0$. The boundary values $y(x_1) = y_1$ and $y(x_2) = y_2$ are given. We can, without loss of generality, assume that $x_2 > x_1$ and $y_2 > y_1$. What is the value of the radius $y(x)$ in between x_1 and x_2 that will minimize the area of this surface?

The area of a thin slice between x and $x + dx$ is $2\pi y(x)ds$ where $ds = \sqrt{1 + y'^2}dx$ is the arc length of the cross-section. Thus the quantity to be minimized is

$$S = \int_{x_1}^{x_2} y(x)\sqrt{1 + y'^2}dx$$

This is the area divided by 2π.

We can derive the Euler–Lagrange equation as before: y is analogous to q and x is analogous to t. But it is smarter to exploit the fact that the integrand is independent of x: there is a conserved quantity

$$H = y'\frac{\partial L}{\partial y'} - L. \quad L = y(x)\sqrt{1 + y'^2}$$

That is

$$H = y\frac{y'^2}{\sqrt{1 + y'^2}} - y\sqrt{1 + y'^2}$$

$$H\sqrt{1 + y'^2} = -y$$

$$y' = \sqrt{\frac{y^2}{H^2} - 1}$$

$$\int_{y_1}^{y} \frac{dy}{\sqrt{\frac{y^2}{H^2} - 1}} = x - x_1$$

The substitution

$$y = H\cosh\theta$$

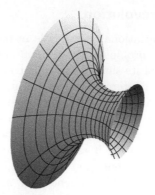

Fig. 2.1 The catenoid.

evaluates the integral:

$$H[\theta - \theta_1] = x - x_1$$

$$\theta = \frac{x - x_1}{H} + \theta_1$$

$$y = H \cosh \left[\frac{x - x_1}{H} + \theta_1 \right]$$

The constants of integration are fixed by the boundary conditions

$$y_1 = H \cosh \theta_1$$

$$y_2 = H \cosh \left[\frac{x_2 - x_1}{H} + \theta_1 \right]$$

The curve $y = H \cosh \left[\frac{x}{H} + \text{constant} \right]$ is called a catenary; the surface you get by revolving it around the x-axis is the catenoid (see Fig. 2.1). If we keep the radii fixed and move the boundaries far apart along the x-axis, at some critical distance, the surface will cease to be of minimal area. The minimal area is given by the disconnected union of two disks with the circles as boundaries.

Problem 2.1: The Lagrangian of the Kepler problem (Example 1.1) does not depend on the angle. What is the conserved quantity implied by this fact?

Problem 2.2: A soap bubble is bounded by two circles of equal radii. If the bounding circles are moved apart slowly, at some distance the bubble will break into two flat disks. Find this critical distance in terms of the bounding radius.

3
The simple pendulum

Consider a mass g suspended from a fixed point by a rigid rod of length l. Also, it is only allowed to move in a fixed vertical plane.

The angle θ from the lowest point on its orbit serves as a position co-ordinate. The kinetic energy is

$$T = \frac{1}{2}ml^2\dot\theta^2$$

and the potential energy is

$$V(\theta) = mgl(1 - \cos\theta)$$

Thus

$$T - V = ml^2\left[\frac{1}{2}\dot\theta^2 - \frac{g}{l}(1 - \cos\theta)\right]$$

The overall constant will not matter to the equations of motion. So we can choose as Lagrangian

$$L = \frac{1}{2}\dot\theta^2 - \frac{g}{l}(1 - \cos\theta)$$

This leads to the equation of motion

$$\ddot\theta + \frac{g}{l}\sin\theta = 0$$

For small angles $\theta \ll \pi$ this is the equation for a harmonic oscillator with angular frequency

$$\omega = \sqrt{\frac{g}{l}}$$

But for large amplitudes of oscillation the answer is quite different. To simplify calculations let us choose a unit of time such that, $g = l$; i.e., such that $\omega = 1$. Then

$$L = \frac{1}{2}\dot\theta^2 - (1 - \cos\theta)$$

We can make progress in solving this system using the conservation of energy

$$H = \frac{\dot{\theta}^2}{2} + [1 - \cos\theta]$$

The key is to understand the critical points of the potential. The potential energy has a minimum at $\theta = 0$ and a maximum at $\theta = \pi$. The latter corresponds to an unstable equilibrium point: the pendulum standing on its head. If the energy is less than this maximum value

$$H < 2$$

the pendulum oscillates back and forth around its equilibrium point. At the maximum angle, $\dot{\theta} = 0$ so that it is given by a transcendental equation

$$1 - \cos\theta_0 = H$$

The motion is periodic, with a period T that depends on energy. That is, we have

$$\sin\theta(t + T) = \sin\theta(t)$$

3.1.   Algebraic formulation

It will be useful to use a variable which takes some simple value at the maximum deflection; also we would like it to be a periodic function of the angle. The condition for maximum deflection can be written

$$\sqrt{\frac{2}{H}}\sin\frac{\theta_0}{2} = \pm 1$$

This suggests that we use the variable

$$x = \sqrt{\frac{2}{H}}\sin\frac{\theta}{2}$$

so that, at maximum deflection, we simply have $x = \pm 1$. Define also a quantity that parametrizes the energy

$$k = \sqrt{\frac{H}{2}}, \quad x = \frac{1}{k}\sin\frac{\theta}{2}$$

Changing variables,

$$\dot{x} = \frac{1}{2k}\cos\frac{\theta}{2}\dot{\theta}, \quad \dot{x}^2 = \frac{1}{4k^2}\left(1 - \sin^2\frac{\theta}{2}\right)\dot{\theta}^2 = \frac{1}{4}\left(\frac{1}{k^2} - x^2\right)\dot{\theta}^2$$

Conservation of energy becomes

$$2k^2 = 2\frac{\dot{x}^2}{k^{-2} - x^2} + 2k^2 x^2$$

Thus we get the differential equation

$$\dot{x}^2 = (1 - x^2)(1 - k^2 x^2)$$

This can be solved in terms of Jacobi functions, which generalize trigonometric functions such as sin and cos.

3.2. Primer on Jacobi functions

The functions $\mathrm{sn}(u, k), \mathrm{cn}(u, k), \mathrm{dn}(u, k)$ are defined as the solutions of the coupled ordinary differential equation (ODE)

$$\mathrm{sn}' = \mathrm{cn}\ \mathrm{dn}, \quad \mathrm{cn}' = -\mathrm{sn}\ \mathrm{dn}, \quad \mathrm{dn}' = -k^2 \mathrm{sn}\ \mathrm{cn}$$

with initial conditions

$$\mathrm{sn} = 0, \quad \mathrm{cn} = 1, \quad \mathrm{dn} = 1, \quad \text{at } u = 0$$

It follows that

$$\mathrm{sn}^2 + \mathrm{cn}^2 = 1, \quad k^2 \mathrm{sn}^2 + \mathrm{dn}^2 = 1$$

Thus

$$\mathrm{sn}'^2 = (1 - \mathrm{sn}^2)(1 - k^2 \mathrm{sn}^2)$$

Thus we see that

$$x(t) = \mathrm{sn}(t, k)$$

is the solution to the pendulum. The inverse of this function (which expresses t as a function of x) can be expressed as an integral

$$t = \int_0^x \frac{dy}{\sqrt{(1 - y^2)(1 - k^2 y^2)}}$$

This kind of integral first appeared when people tried to find the perimeter of an ellipse. So it is called an *elliptic integral*.

The functions sn, cn, dn are called *elliptic functions*. The name is a bit unfortunate, because these functions appear even when there is no ellipse in sight, such as in our case. The parameter k is called the *elliptic modulus*.

Clearly, if $k = 0$, these functions reduce to trigonometric functions:

$$\text{sn}(u,0) = \sin u, \quad \text{cn}(u,0) = \cos u, \quad \text{dn}(u,0) = 1$$

Thus, for small energies $k \to 0$ and our solution reduces to that of the harmonic oscillator.

From the connection with the pendulum it is clear that the functions are periodic, at least when $0 < k < 1$ (so that $0 < H < 2$ and the pendulum oscillates around the equilibrium point). The period of oscillation is four times the time it takes to go from the bottom to the point of maximum deflection

$$T = 4K(k), \quad K(k) = \int_0^1 \frac{dy}{\sqrt{(1-y^2)(1-k^2y^2)}}$$

This integral is called the *complete elliptic integral*. When $k = 0$, it evaluates to $\frac{\pi}{2}$ so that the period is 2π. That is correct, since we chose the unit of time such that $\omega = \sqrt{\frac{l}{g}} = 1$ and the period of the harmonic oscillator is $\frac{2\pi}{\omega}$. As k grows, the period increases: the pendulum oscillates with larger amplitude. As $k \to 1$ the period tends to infinity: the pendulum has just enough energy to get to the top of the circle, with velocity going to zero as it gets there.

3.3. Elliptic curves*

Given the position x and velocity $\dot{x}$ at any instant, they are determined for all future times by the equations of motion. Thus it is convenient to think of a space whose co-ordinates are $(x, \dot{x})$. The conservation of energy determines the shape of the orbit in phase space.

$$\dot{x}^2 = (1-x^2)(1-k^2x^2)$$

In the case of a pendulum, this is an extremely interesting thing called an *elliptic curve*. The first thing to know is that *an elliptic curve is not an ellipse*. It is called that because elliptic functions can be used to parametrically describe points on this curve:

$$\dot{x} = \text{sn}'(u,k), \quad x = \text{sn}(u,k)$$

For small k an elliptic curve looks more or less like a circle, but as $k > 0$ it is deformed into a more interesting shape. When $k \to 1$ it tends to a parabola (see Fig. 3.1).

Only the part of the curve with real x between 1 and -1 has a physical significance in this application. But, as usual, to understand any algebraic curve it helps to analytically continue into the complex plane. The surprising thing is that the curve is then a torus; this follows from the double periodicity of sn, which we prove below.

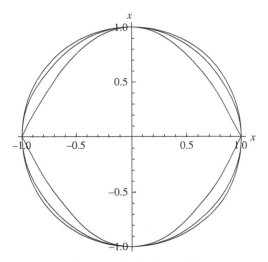

Fig. 3.1 Elliptic curves.

3.3.1   Addition formula

Suppose x_1 is the position of the pendulum after a time t_1 has elapsed, assuming that at time zero, $x = 0$ as well. Similarly, let x_2 be the position at some time t_2. If $t_3 = t_1 + t_2$, and x_3 is the position at time t_3, it should not be surprising that x_3 can be found once we know x_1 and x_2. What is surprising is that x_3 is an *algebraic* function of the positions. This is because of the addition formula for elliptic functions:

$$\int_0^{x_1} \frac{dy}{\sqrt{(1-y^2)(1-k^2y^2)}} + \int_0^{x_2} \frac{dy}{\sqrt{(1-y^2)(1-k^2y^2)}} = \int_0^{x_3} \frac{dy}{\sqrt{(1-y^2)(1-k^2y^2)}}$$

$$x_3 = \frac{x_1\sqrt{(1-x_2^2)(1-k^2x_2^2)} + x_2\sqrt{(1-x_1^2)(1-k^2x_1^2)}}{1 - k^2x_1^2x_2^2}$$

If $k = 0$, this is just the addition formula for sines:

$$\sin[\theta_1 + \theta_2] = \sin\theta_1 \sqrt{1 - \sin^2\theta_2} + \sin\theta_2 \sqrt{1 - \sin^2\theta_1}.$$

This operation $x_1, x_2 \mapsto x_3$ satisfies the conditions for an abelian group. The point of stable equilibrium $x = 0, \dot{x} = 1$ is the identity element. The inverse of x_1 is just $-x_1$. You can have fun trying to prove algebraically that the above operation is associative. (Or die trying.)

3.4.   Imaginary time

The replacement $t \to it$ has the effect of reversing the sign of the potential in Newton's equations, $\ddot{q} = -V'(q)$. In our case, $\ddot{\theta} = -\sin\theta$, it amounts to reversing the direction of the gravitational field. In terms of co-ordinates, this amounts to $\theta \to \theta + \pi$. Under the transformations $t \to it, \theta \to \theta + \pi$, the conservation of energy

$$k^2 = \frac{\dot{\theta}^2}{4} + \sin^2\frac{\theta}{2}$$

goes over to

$$1 - k^2 = \frac{\dot{\theta}^2}{4} + \sin^2\frac{\theta}{2}$$

The quantity $k' = \sqrt{1 - k^2}$ is called the *complementary modulus*. In summary, the simple pendulum has a symmetry

$$t \to it, \theta \to \theta + \pi, k \to k'$$

This transformation maps an oscillation of small amplitude (small k) to one of large amplitude (k close to 1).

This means that if we analytically continue the solution of the pendulum into the complex t-plane, it must be periodic with period $4K(k)$ in the real direction and $4K(k')$ in the imaginary direction.

3.4.1   The Case of $H = 1$

The minimum value of energy is zero and the maximum value for an oscillation is 2. Exactly half way is the oscillation whose energy is 1; the maximum angle is $\frac{\pi}{2}$. This orbit is invariant under the above transformation that inverts the potential: either way you look at it, the pendulum bob is horizontal at maximum deflection. In this case, the real and imaginary periods are of equal magnitude.

3.5.   The arithmetic-geometric mean*

Landen, and later Gauss, found a surprising symmetry for the elliptic integral $K(k)$ that allows a calculation of its value by iterating simple algebraic operations. In our context it means that the period of a pendulum is unchanged if the energy H and angular frequency ω are changed in a certain way that decreases their values. By iterating this we can make the energy tend to zero, but in this limit we know that the period is just 2π over the angular frequency. *In this section we do not set $\omega = 1$ but we continue to factor out ml^2 from the Lagrangian as before.* Then the Lagrangian $L = \frac{1}{2}\dot{\theta}^2 - \omega^2[1 - \cos\theta]$ and H have dimensions of the square of frequency.

Let us go back and look at the formula for the period:

$$T = \frac{4}{\omega}K(k), \quad K(k) = \int_0^1 \frac{dx}{\sqrt{(1 - x^2)(1 - k^2 x^2)}}, \quad 2k^2 = \frac{H}{\omega^2}$$

If we make the substitution

$$x = \sin\phi$$

this becomes

$$T = \frac{4}{\omega} \int_0^{\frac{\pi}{2}} \frac{d\phi}{\sqrt{1 - k^2 \sin^2 \phi}}$$

That is,

$$T(\omega, b) = 4 \int_0^{\frac{\pi}{2}} \frac{d\phi}{\sqrt{\omega^2 \cos^2 \phi + b^2 \sin^2 \phi}}$$

where

$$b = \sqrt{\omega^2 - \frac{H}{2}}$$

Note that $\omega > b$ with $\omega \to b$ implying $H \to 0$. The surprising fact is that the integral remains unchanged under the transformations

$$\omega_1 = \frac{\omega + b}{2}, \quad b_1 = \sqrt{\omega b}$$

$$T(\omega, b) = T(\omega_1, b_1)$$

Exercise 3.1: Prove this identity. First put $y = b \tan \phi$ to get $T(\omega, b) = 2 \int_{-\infty}^{\infty} \frac{dy}{\sqrt{(\omega^2 + y^2)(b^2 + y^2)}}$. Then make the change of variable $y = z + \sqrt{z^2 + \omega b}$. This proof, due to Newman, was only found in 1985. Gauss's and Landen's proofs were much clumsier. For further explanation, see McKean and Moll (1999).

That is, ω is replaced by the arithmetic mean and b by the geometric mean. Recall that given two numbers $a > b > 0$, the arithmetic mean is defined by

$$a_1 = \frac{a + b}{2}$$

and the geometric mean is defined as

$$b_1 = \sqrt{ab}$$

As an exercise it is easy to prove that in general $a_1 \geq b_1$. If we iterate this transformation,

$$a_{n+1} = \frac{a_n + b_n}{2}, \quad b_{n+1} = \sqrt{a_n b_n}$$

the two sequences converge to the same number, $a_n \to b_n$ as $n \to \infty$. This limiting value is called the *Arithmetic-Geometric Mean* AGM(a, b).

Thus, the energy of the pendulum tends to zero under this iteration applied to ω and b, since $\omega_n \to b_n$; and the period is the limit of $\frac{2\pi}{\omega_n}$:

$$T(\omega, b) = \frac{2\pi}{\text{AGM}(\omega, b)}$$

The convergence of the sequence is quite fast, and gives a very accurate and elementary way to calculate the period of a pendulum, i.e., without having to calculate any integral.

3.5.1   The arithmetic-harmonic mean is the geometric mean

Why would Gauss have thought of the arithmetic-geometric mean? This is perhaps puzzling to a modern reader brought up on calculators. But it is not so strange if you know how to calculate square roots by hand.

Recall that the harmonic mean of a pair of numbers is the reciprocal of the mean of their reciprocals. That is

$$\mathrm{HM}(a,b) = \frac{1}{\frac{1}{2}\left(\frac{1}{a} + \frac{1}{b}\right)} = \frac{2ab}{a+b}$$

Using $(a+b)^2 > 4ab$, it follows that $\frac{a+b}{2} > \mathrm{HM}(a,b)$. Suppose that we define an iterative process whereby we take the successive arithmetic and harmonic means:

$$a_{n+1} = \frac{a_n + b_n}{2}, \quad b_{n+1} = \mathrm{HM}(a_n, b_n)$$

These two sequences approach each other, and the limit can be defined to be the arithmetic-harmonic mean (AHM).

$$\mathrm{AHM}(a,b) = \lim_{n\to\infty} a_n$$

In other words, $\mathrm{AHM}(a,b)$ is defined by the invariance properties

$$\mathrm{AHM}(a,b) = \mathrm{AHM}\left(\frac{a+b}{2}, \mathrm{HM}(a,b)\right), \quad \mathrm{AHM}(\lambda a, \lambda b) = \lambda \mathrm{AHM}(a,b)$$

What is this quantity? It is none other than the geometric mean! Simply verify that

$$\sqrt{ab} = \sqrt{\frac{a+b}{2}\frac{2ab}{a+b}}$$

Thus iterating the arithmetic and harmonic means with 1 is a good way to calculate the square root of any number. (Try it.)

Now you see that it is natural to wonder what we would get if we do the same thing one more time, iterating the arithmetic and the geometric means.

$$\mathrm{AGM}(a,b) = \mathrm{AGM}\left(\frac{a+b}{2}, \sqrt{ab}\right)$$

I don't know if this is how Gauss discovered it, but it is not such a strange idea.

Exercise 3.2: Relate the harmonic-geometric mean, defined by the invariance below to the AGM.

$$\mathrm{HGM}(a,b) = \mathrm{HGM}\left(\frac{2ab}{a+b}, \sqrt{ab}\right)$$

3.6. Doubly periodic functions*

We are led to the modern definition: *an elliptic function is a doubly periodic analytic function of a complex variable.* We allow for poles, but not branch cuts: thus, to be precise, an elliptic function is a doubly periodic meromorphic function of a complex variable.

$$f(z + m_1\tau_1 + m_2\tau_2) = f(z)$$

for integer m_1, m_2 and complex numbers τ_1, τ_2 which are called the periods. The points at which f takes the same value form a lattice in the complex plane. Once we know the values of f in a parallelogram whose sides are τ_1 and τ_2, we will know it everywhere by translation by some integer linear combination of the periods. In the case of the simple pendulum above, one of the periods is real and the other is purely imaginary. More generally, they could both be complex numbers; as long as the area of the fundamental parallelogram is non-zero, we will get a lattice. By a rotation and a rescaling of the variable, we can always choose one of the periods to be real. The ratio of the two periods

$$\tau = \frac{\tau_2}{\tau_1}$$

is thus the quantity that determines the shape of the lattice. It is possible to take some rational function and sum over its values at the points $z + m_1\tau_1 + m_2\tau_2$ to get a doubly periodic function, provided that this sum converges. An example is

$$\mathcal{P}'(z) = -2 \sum_{m_1,m_2=-\infty}^{\infty} \frac{1}{(z + m_1\tau_1 + m_2\tau_2)^3}$$

The power 3 in the denominator is the smallest one for which this sum converges; the factor of -2 in front is there to agree with some conventions. It has triple poles at the origin and all points obtained by translation by periods $m_1\tau_1 + m_2\tau_2$. It is the derivative of another elliptic function called $\mathcal{P}$, the *Weierstrass elliptic function*. It is possible to express the Jacobi elliptic functions in terms of the Weierstrass function; these two approaches complement each other. See McKean and Moll (1999) for more on these matters.

Problem 3.3: Show that the perimeter of the ellipse $\frac{x^2}{a^2} + \frac{y^2}{b^2} = 1$ is equal to $4a \int_0^1 \sqrt{\frac{1-k^2x^2}{1-x^2}} dx$, where $k = \sqrt{1 - \frac{b^2}{a^2}}$.

Problem 3.4: Using the change of variable $x \mapsto \frac{1}{\sqrt{1-k'^2x^2}}$, show that $K(k') = \int_1^{\frac{1}{k}} \frac{dx}{\sqrt{[x^2-1][1-k^2x^2]}}$.

Problem 3.5: Calculate the period of the pendulum with $\omega = 1, H = 1$ by calculating the arithmetic-geometric mean. How many iterations do you need to get an accuracy of five decimal places for the AGM?

Problem 3.6**: What can you find out about the function defined by the following invariance property?

$$\mathrm{AAGM}(a, b) = \mathrm{AAGM}\left(\frac{a+b}{2}, \mathrm{AGM}(a, b)\right)$$

10.1 Doubly periodic functions

4
The Kepler problem

Much of mechanics was developed in order to understand the motion of planets. Long before Copernicus, many astronomers knew that the apparently erratic motion of the planets can be simply explained as circular motion around the Sun. For example, the *Aryabhateeyam* written in AD 499 gives many calculations based on this model. But various religious taboos and superstitions prevented this simple picture from being universally accepted. It is ironic that the same superstitions (e.g., astrology) were the prime cultural motivation for studying planetary motion. Kepler himself is a transitional figure. He was originally motivated by astrology, yet had the scientific sense to insist on precise agreement between theory and observation.

Kepler used Tycho Brahe's accurate measurements of planetary positions to find a set of important refinements of the heliocentric model. The three laws of planetary motion that he discovered started the scientific revolution which is still continuing. We will rearrange the order of presentation of the laws of Kepler to make the logic clearer. Facts are not always discovered in the correct logical order: reordering them is essential to understanding them.

4.1. The orbit of a planet lies on a plane which contains the Sun

We may call this the zeroth law of planetary motion: this is a significant fact in itself. If the *direction* of angular momentum is preserved, the orbit would have to lie in a plane. Since $\mathbf{L} = \mathbf{r} \times \mathbf{p}$, this plane is normal to the direction of $\mathbf{L}$. In polar co-ordinates in this plane, the angular momentum is

$$L = mr^2 \frac{d\phi}{dt}$$

that is, the moment of inertia times the angular velocity. In fact, all the planetary orbits lie on the *same* plane to a good approximation. This plane is normal to the angular momentum of the original gas cloud that formed the solar system.

4.2. The line connecting the planet to the Sun sweeps equal areas in equal times

This is usually called the second law of Kepler. Since the rate of change of this area is $\frac{r^2}{2} \frac{d\phi}{dt}$, this is the statement that

$$r^2 \frac{d\phi}{dt} = \text{constant}$$

This can be understood as due to the conservation of angular momentum. If the force is always directed towards the Sun, this can be explained.

4.3.   Planets move along elliptical orbits with the Sun at a focus

This is the famous first law of Kepler. It is significant that the orbit is a closed curve and that it is periodic: for most central potentials neither statement is true.

An *ellipse* is a curve on the plane defined by the equation, in polar co-ordinates r, ϕ

$$\frac{\rho}{r} = 1 + \epsilon \cos \phi$$

For an ellipse, the parameter ϵ must be between 0 and 1 and is called the *eccentricity*. It measures the deviation of an ellipse from a circle: if $\epsilon = 0$ the curve is a circle of radius ρ. In the opposite limit $\epsilon \to 1$ (keeping ρ fixed) it approaches a parabola. The parameter ρ measures the size of the ellipse.

A more geometrical description of the ellipse is this: Choose a pair of points on the plane F_1, F_2, the *focii*. If we let a point move on the plane such that the sum of its distances to F_1 and F_2 is a constant, it will trace out an ellipse.

Exercise 4.1: Derive the equation for the ellipse above from this geometrical description. (Choose the origin of the polar co-ordinate system to be F_1. What is the position of the other focus?)

The line connecting the two farthest points on an ellipse is called its *major axis*; this axis passes through the focii. The perpendicular bisector to the major axis is the *minor axis*. If these are equal in length, the ellipse is a circle; in this case the focii coincide. Half of the length of the major axis is called a usually. Similarly, the semi-minor axis is called b.

Exercise 4.2: Show that the major axis is $\frac{2\rho}{1-\epsilon^2}$ and that the eccentricity is $\epsilon = \sqrt{1 - \frac{b^2}{a^2}}$.

The eccentricity of planetary orbits is quite small: a few percent. Comets, some asteroids and planetary probes have very eccentric orbits. If the eccentricity is greater than one, the equation describes a curve that is not closed, called a *hyperbola*. In the *Principia*, Newton proved that an elliptical orbit can be explained by a force directed towards the Sun that is inversely proportional to the square of distance. Where did he get the idea of a force inversely proportional to the square of distance? The third law of Kepler provides a clue.

4.4.   The ratio of the cube of the semi-major axis to the square of the period is the same for all planets

It took 17 years of hard work for Kepler to go from the second Law to this third law. Along the way, he considered and discarded many ideas on planetary distances that came from astrology and Euclidean geometry (Platonic solids).

If, for the moment, we ignore the eccentricity (which is anyway small) and consider just a circular orbit of radius r, this is saying that

$$T^2 \propto r^3$$

We already know that the force on the planet must be pointed toward the Sun, from the conservation of angular momentum. What is the dependence of the force on distance that will give this dependence of the period? Relating the force to the centripetal acceleration,

$$m\frac{v^2}{r} = F(r)$$

Now, $v = r\dot{\theta}$ and $\dot{\theta} = \frac{2\pi}{T}$ for uniform circular motion. Thus

$$T^2 \propto \frac{r}{F(r)}$$

So we see that $F(r) \propto \frac{1}{r^2}$.

Hooke, a much less renowned scientist than Newton, verified using a mechanical model that orbits of particles in this force are ellipses. He made the suggestion to Newton, who did not agree at that time. Later, while he was writing the *Principia*, Newton discovered a marvelous proof of this fact using only geometry (no calculus). Discoveries are often made by a collaborative process involving many people, not just a lone genius.

From the fact that the ratio $\frac{T^2}{r^3}$ is independent of the planet, we can conclude that the acceleration is independent of the mass of the planet: that the force is proportional to the product of masses. Thus we arrive at Newton's Law of Gravity:

> The gravitational force on a body due to another is pointed along the line connecting the bodies; it has magnitude proportional to the product of masses and inversely to the square of the distance.

4.5.   The shape of the orbit

We now turn to deriving the shape of a planetary orbit from Newton's law of gravity. The Lagrangian is, in plane polar co-ordinates centered at the Sun,

$$L = \frac{1}{2}m\left[\dot{r}^2 + r^2\dot{\phi}^2\right] + \frac{GMm}{r}$$

From this we deduce the momenta

$$p_r = m\dot{r}, \quad p_\phi = mr^2\dot{\phi}$$

and the hamiltonian

$$H = \frac{p_r^2}{2m} + \frac{p_\phi^2}{2mr^2} - \frac{GMm}{r}$$

Since $\frac{\partial H}{\partial \phi} = 0$, it follows right away that p_ϕ is conserved.

$$H = \frac{p_r^2}{2m} + V(r)$$

where

$$V(r) = \frac{p_\phi^2}{2mr^2} - \frac{GMm}{r}$$

is an *effective potential*, the sum of the gravitational potential and the kinetic energy due to angular motion (see Fig. 4.1).

So,

$$\dot{r} = \frac{p_r}{m}$$
$$\dot{p_r} = -V'(r)$$

Right away, we see that there is a circular orbit at the minimum of the potential:

$$V'(r) = 0 \implies r = \frac{p_\phi^2}{GMm^2}$$

More generally, when $H < 0$, we should expect an oscillation around this minimum, between the *turning points*, which are the roots of $H - V(r) = 0$. For $H > 0$ the particle will come in from infinity and, after reflection at a turning point, escape back to infinity.

The shape of the orbit is given by relating r to ϕ. Using

$$\frac{dr}{dt} = \frac{d\phi}{dt}\frac{dr}{d\phi} = \frac{p_\phi}{mr^2}\frac{dr}{d\phi}$$

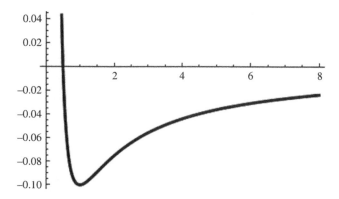

Fig. 4.1 The potential of the Kepler problem.

This suggests the change of variable

$$u = A + \frac{\rho}{r}, \implies \frac{dr}{dt} = -\frac{p_\phi}{m\rho}\frac{du}{d\phi} = -\frac{p_\phi}{m\rho}u'$$

for some constants A, ρ that we will choose for convenience later. We can express the conservation of energy

$$H = \frac{1}{2}m\dot{r}^2 + V(r)$$

as

$$H = \frac{p_\phi^2}{2m\rho^2}u'^2 + \frac{p_\phi^2}{2m\rho^2}(u - A)^2 - \frac{GMm}{\rho}(u - A)$$

$$\frac{2m\rho^2 H}{p_\phi^2} = u'^2 + (u - A)^2 - \frac{2GMm^2\rho}{p_\phi^2}(u - A)$$

We can now choose the constants so that the term linear in u cancels out

$$A = -1, \quad \rho = \frac{p_\phi^2}{GMm^2}$$

and

$$u'^2 + u^2 = \epsilon^2$$

$$\epsilon^2 = 1 + \frac{2p_\phi^2 H}{(GM)^2 m^3}$$

A solution is now clear

$$u = \epsilon \cos\phi$$

or

$$\frac{\rho}{r} = 1 + \epsilon \cos\phi$$

This is the equation for a conic section of eccentricity ϵ. If $H < 0$, so that the planet cannot escape to infinity, this is less than one, giving an ellipse as the orbit.

Problem 4.3: Show that among all Kepler orbits of the same angular momentum, the circle has the least energy.

Problem 4.4: What would be the shape of the orbit if the gravitational potential had a small correction that varies inversely with the square of the distance? Which of the laws of planetary motion would still hold?

Problem 4.5: The equation of motion of a classical electron orbiting a nucleus is, including the effect of radiation,

$$\frac{d}{dt}\left[\dot{\mathbf{r}} + \tau \nabla U\right] + \nabla U = 0, \quad U = -\frac{k}{r}$$

The positive constants τ, k are related to the charge and mass of the particles. Show that the orbits are spirals converging to the center. This problem was solved in Rajeev (2008).

5
The rigid body

<hr/>

If the distance between any two points on a body remains constant as it moves, it is a rigid body. Any configuration of the rigid body can be reached from the initial one by a translation of its center of mass and a rotation around it. Since we are mostly interested in the rotational motion, we will only consider the case of a body on which the total force is zero: the center of mass moves at a constant velocity. In this case we can transform to the reference frame in which the center of mass is at rest: the origin of our co-ordinate system can be placed there. It is not hard to put back in the translational degree of freedom once rotations are understood.

The velocity of one of the particles making up the rigid body can be split as

$$\mathbf{v} = \mathbf{\Omega} \times \mathbf{r}$$

The vector $\mathbf{\Omega}$ is the *angular velocity*: its direction is the axis of rotation and its magnitude is the rate of change of its angle. The kinetic energy of this particle inside the body is

$$\frac{1}{2} \left[\mathbf{\Omega} \times \mathbf{r} \right]^2 \rho(\mathbf{r}) d^3 \mathbf{r}$$

Here $\rho(\mathbf{r})$ is the mass density at the position of the particle; we assume that it occupies some infinitesimally small volume $d^3\mathbf{r}$. Thus the total rotational kinetic energy is

$$T = \frac{1}{2} \int \left[\mathbf{\Omega} \times \mathbf{r} \right]^2 \rho(\mathbf{r}) d^3 \mathbf{r}$$

Now, $(\mathbf{\Omega} \times \mathbf{r})^2 = \Omega^2 r^2 - (\mathbf{\Omega} \cdot \mathbf{r})^2 = \Omega_i \Omega_j \left[r^2 \delta_{ij} - r_i r_j \right]$, so we get

$$T = \frac{1}{2} \Omega_i \Omega_j \int \rho(\mathbf{r}) \left[r^2 \delta_{ij} - r_i r_j \right] d^3 \mathbf{r}$$

5.1. The moment of inertia

Define the *moment of inertia* to be the symmetric matrix

$$I_{ij} = \int \rho(\mathbf{r}) \left[r^2 \delta_{ij} - r_i r_j \right] d^3 \mathbf{r}$$

Thus

$$T = \frac{1}{2}\Omega_i\Omega_j I_{ij}$$

Being a symmetric matrix, there is an orthogonal co-ordinate system in which the moment of inertia is diagonal:

$$T = \frac{1}{2}\left[I_1\Omega_1^2 + I_2\Omega_2^2 + I_3\Omega_3^2\right]$$

The eigenvalues I_1, I_2, I_3 are called the *principal moments of inertia*. They are positive numbers because I_{ij} is a positive matrix, i.e., $u^T I u \geq 0$ for any u.

> **Exercise 5.1:** Show that the sum of any two principal moments is greater than or equal to the third one. $I_1 + I_2 \geq I_3$ etc.

The shape of the body, and how mass is distributed inside it, determines the moment of inertia. The simplest case is when all three are equal. This happens if the body is highly symmetric: a sphere, a regular solid such as a cube. The next simplest case is when two of the moments are equal and the third is different. This is a body that has one axis of symmetry: a cylinder, a prism whose base is a regular polygon etc. The most complicated case is when the three eigenvalues are all unequal. This is the case of the asymmetrical top.

5.2.   Angular momentum

The angular momentum of a small particle inside the rigid body is

$$dM\mathbf{r} \times \mathbf{v} = \left[\rho(\mathbf{r})d^3\mathbf{r}\right]\mathbf{r} \times (\mathbf{\Omega} \times \mathbf{r})$$

Using the identity $\mathbf{r} \times (\mathbf{\Omega} \times \mathbf{r}) = \mathbf{\Omega}r^2 - \mathbf{r}(\mathbf{\Omega} \cdot \mathbf{r})$ we get the total angular momentum of the body to be

$$\mathbf{L} = \int \rho(\mathbf{r})[r^2\mathbf{\Omega} - \mathbf{r}(\mathbf{\Omega} \cdot \mathbf{r})]d^3\mathbf{r}$$

In terms of components

$$L_i = I_{ij}\Omega_j$$

Thus the moment of inertia relates angular velocity to angular momentum, just as mass relates velocity to momentum. The important difference is that moment of inertia is a matrix so that $\mathbf{L}$ and $\mathbf{\Omega}$ do not have to point in the same direction. Recall that the rate of change of angular momentum is the torque, if they are measured in an inertial reference frame.

Now here is a tricky point. We would like to use a co-ordinate system in which the moment of inertia is a diagonal matrix; that would simplify the relation of angular momentum to angular velocity:

$$L_1 = I_1 \Omega_1$$

etc. But this may not be an inertial co-ordinate system, as its axes have to rotate with the body. So we must relate the change of a vector (such as $\mathbf{L}$) in a frame that is fixed to the body to an inertial frame. The difference between the two is a rotation of the body itself, so that

$$\left[\frac{d\mathbf{L}}{dt}\right]_{\text{inertial}} = \frac{d\mathbf{L}}{dt} + \mathbf{\Omega} \times \mathbf{L}$$

This we set equal to the torque acting on the body as a whole.

5.3.   Euler's equations

Even in the special case when the torque is zero the equations of motion of a rigid body are non-linear, since $\mathbf{\Omega}$ and $\mathbf{L}$ are proportional to each other:

$$\frac{d\mathbf{L}}{dt} + \mathbf{\Omega} \times \mathbf{L} = 0$$

In the co-ordinate system with diagonal moment of inertia

$$\Omega_1 = \frac{L_1}{I_1}$$

these become

$$\frac{dL_1}{dt} + a_1 L_2 L_3 = 0, \quad a_1 = \frac{1}{I_2} - \frac{1}{I_3}$$

$$\frac{dL_2}{dt} + a_2 L_3 L_1 = 0, \quad a_2 = \frac{1}{I_3} - \frac{1}{I_1}$$

$$\frac{dL_3}{dt} + a_3 L_1 L_2 = 0, \quad a_3 = \frac{1}{I_1} - \frac{1}{I_2}$$

These equations were originally derived by Euler. Clearly, if all the principal moments of inertia are equal, these are trivial to solve: $\mathbf{L}$ is a constant.

The next simplest case

$$I_1 = I_2 \neq I_3$$

is not too hard either. Then $a_3 = 0$ and $a_1 = -a_2$.

It follows that L_3 is a constant. Also, L_1 and L_2 *precess* around this axis:

$$L_1 = A\cos\omega t, \quad L_2 = A\sin\omega t$$

with

$$\omega = a_1 L_3$$

An example of such a body is the Earth. It is not quite a sphere, because it bulges at the equator compared to the poles. The main motion of the Earth is its rotation around the north–south axis once every 24 hours. But this axis itself precesses once every 26,000 years. This means that the axis was not always aligned with the Pole Star in the distant past. Also, the times of the equinoxes change by a few minutes each year. As early as 280 BC Aristarchus described this precession of the equinoxes. It was Newton who finally explained it physically.

5.4.   Jacobi's solution

The general case of unequal moments can be solved in terms of Jacobi elliptic functions; in fact, these functions were invented for this purpose. But before we do that it is useful to find the constants of motion. It is no surprise that the energy

$$H = \frac{1}{2}I_1\Omega_1^2 + \frac{1}{2}I_2\Omega_2^2 + \frac{1}{2}I_3\Omega_3^2 = \frac{L_1^2}{2I_1} + \frac{L_2^2}{2I_2} + \frac{L_3^2}{2I_3}$$

is conserved. You can verify that the magnitude of angular momentum is conserved as well:

$$L^2 = L_1^2 + L_2^2 + L_3^2$$

Exercise 5.2: Calculate the time derivatives of H and L^2 and verify that they are zero.

Recall that

$$\text{sn}' = \text{cn dn}, \quad \text{cn}' = -\text{sn dn}, \quad \text{dn}' = -k^2\text{sn cn}$$

with initial conditions

$$\text{sn} = 0, \quad \text{cn} = 1, \quad \text{dn} = 1, \quad \text{at } u = 0$$

Moreover

$$\text{sn}^2 + \text{cn}^2 = 1, \quad k^2\text{sn}^2 + \text{dn}^2 = 1$$

Make the ansatz

$$L_1 = A_1\text{cn}(\omega t, k) \quad L_2 = A_2\text{sn}(\omega t, k), \quad L_3 = A_3\text{dn}(\omega t, k)$$

We get conditions

$$-\omega A_1 + a_1 A_2 A_3 = 0$$
$$\omega A_2 + a_2 A_3 A_1 = 0$$
$$-\omega k^2 A_3 + a_3 A_1 A_2 = 0$$

We want to express the five constants A_1, A_2, A_3, ω, k that appear in the solution in terms of the five physical parameters H, L, I_1, I_2, I_3. Some serious algebra will give[1]

$$\omega = \sqrt{\frac{(I_3 - I_2)(L^2 - 2HI_1)}{I_1 I_2 I_3}}$$

$$k^2 = \frac{(I_2 - I_1)(2HI_3 - L^2)}{(I_3 - I_2)(L^2 - 2HI_1)}$$

and

$$A_2 = \sqrt{\frac{(2HL_3 - L^2)I_2}{I_3 - I_2}}$$

etc.

The quantum mechanics of the rigid body is of much interest in molecular physics. So it is interesting to reformulate this theory in a way that makes the passage to quantum mechanics more natural. The Poisson brackets of angular momentum derived later give such a formulation.

Problem 5.3: Show that the principal moments of inertia of a cube of constant density are all equal. So, there is a sphere of some radius with the same moment of inertia and density as the cube. What is its radius as a multiple of the side of the cube?

Problem 5.4: More generally, show that the moment of inertia is proportional to the identity matrix for all of the regular solids of Euclidean geometry. In addition to the cube, these are the tetrahedron, the octahedron, the dodecahedron and the icosahedron. A little group theory goes a long way here.

Problem 5.5: A spheroid is the shape you get by rotating an ellipse around one of its axes. If it is rotated around the major (minor) axis you get a prolate (oblate) spheroid. Find the principal moments of inertia for each type of spheroid.

---

[1] We can label our axes such that $I_3 > I_2 > I_1$.

6

Geometric theory of ordinary differential equations

Any problem in classical mechanics can be reduced to a system of ordinary differential equations

$$\frac{dx^{\mu}}{dt} = V^{\mu}(x), \quad \text{for } \mu = 1, \cdots, n$$

In general V^{μ} can depend on the independent variable t in addition to x^{μ}. But we will avoid this by a cheap trick: if $V^{\mu}(x,t)$ does depend on time, we will add an extra variable x^{n+1} and an extra component $V^{n+1}(x^1, \cdots, x^{n+1}) = 1$. This says that $x^{n+1} = t$ up to a constant, so we get back the original situation. Similarly, if we have to solve an equation of order higher than one, we can just add additional variables to bring it to this first order form. Also, we will assume in this chapter that V^{μ} are differentiable functions of x^{μ}. Singular cases have to be studied by making a change of variables that regularizes them (i.e., brings them to non-singular form).

For more on this geometric theory, including proofs, study the classic texts (Lefschetz, 1977; Arnold, 1978).

6.1. Phase space

We can regard the dependent variables as the co-ordinates of some space, called the *phase space*.[1] The number of variables n is then the *dimension* of the phase space. Given the variables x^{μ} at one instant of the independent variable ("time"), it tells us how to deduce their values a small time later: $x^{\mu}(t + dt) = x^{\mu}(t) + V^{\mu}(x)dt$. It is useful to have a physical picture: imagine a fluid filling the phase space, and $V^{\mu}(x)$ is the velocity of the fluid element at x^{μ}. At the next instant it has moved to a new point where there is a new value of the velocity, which then tells it where to be at the instant after and so on. Thus, V^{μ} defines a vector field in phase space. Through every point in phase spaces passes a curve which describes the time evolution of that point: it is called the integral curve of the vector field V^{μ}. These curves intersect only at a point where the vector field vanishes: these are *fixed points*.

---

[1] By "space" we mean "differential manifold." It is a tradition in geometry to label co-ordinates by an index that sits above (superscript) instead of a subscript. You can tell by the context whether x^2 refers to the second component of some co-ordinate or the square of some variable x.

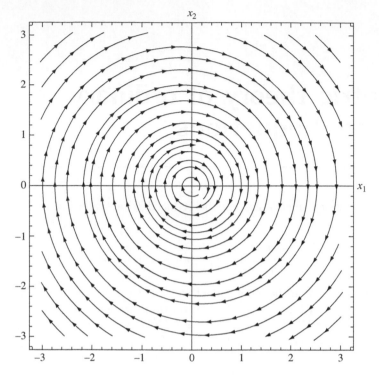

Fig. 6.1 Orbits of the damped harmonic oscillator.

Example 6.1: The equation of a damped simple harmonic oscillator

$$\ddot{q} + \gamma\dot{q} + \omega^2 q = 0$$

is reduced to first order form by setting $x^1 = q, x^2 = \dot{q}$ so that

$$\frac{dx^1}{dt} = x^2, \quad \frac{dx^2}{dt} = -\gamma x^2 - \omega^2 x^1$$

The phase space in this case is the plane. The origin is a fixed point. The integral curves are spirals towards the origin (see Fig. 6.1).

6.2.   Differential manifolds

The simplest example of a differential manifold is Euclidean space. By introducing a Cartesian co-ordinate system, we can associate to each point an ordered tuple of real numbers $(x^1, x^2, \cdots, x^n)$, its co-ordinates. Any function $f : M \to R$ can then be viewed as a function of the co-ordinates. We know what it means for a function of several variables to be smooth, which we can use to define a smooth function in Euclidean space. But there is nothing sacrosanct about a Cartesian co-ordinate system. We can make a change of variables to a new system

$$y^\mu = \phi^\mu(x)$$

as long as the new co-ordinates are still smooth, invertible functions of the old. In particular, this means that the matrix (Jacobian) $J^\mu_\nu = \frac{\partial \phi^\mu}{\partial x^\nu}$ has non-zero determinant everywhere. In fact, if we had the inverse function $x^\mu = \psi^\mu(y)$ its Jacobian would be the inverse:

$$\frac{\partial \phi^\mu}{\partial x^\nu} \frac{\partial \psi^\nu}{\partial y^\sigma} = \delta^\mu_\sigma$$

But it turns out to be too strict to demand that the change of co-ordinates be smooth everywhere: sometimes we want co-ordinates systems that are only valid in some subset of Euclidean space.

Example 6.2: The transformation to the polar co-ordinate system on the plane

$$x^1 = r\cos\theta, \quad x^2 = r\sin\theta$$

is not invertible because θ and $\theta + 2\pi$ correspond to the same point. So we must restrict to the range $0 \le \theta < 2\pi$. Thus the polar system is valid on the plane with the positive x^1-axis removed. To cover all of Euclidean space, we must use, in addition, another polar co-ordinate system with a different origin and an axis that does not intersect this line. Most points will belong to both co-ordinate patches, and the transformation between them is smooth. Together they cover all of the plane.

We can abstract out of this the definition of a far-reaching concept. A differential manifold is a set which can be written as a union of $M = \bigcup_\alpha U_\alpha$; each "co-ordinate patch" U_α is in one–one correspondence $x_\alpha : U_\alpha \to R^n$ with some domain of Euclidean space R^n. Moreover, there are smooth invertible functions ("co-ordinate transformations") $\phi_{\alpha\beta} : R^n \to R^n$ such that

$$x_\alpha = \phi_{\alpha\beta}(x_\beta)$$

The point of this is that locally (in the neighborhood of a point) a differentiable manifold looks like Euclidean space. But each such patch can be attached to another, such that globally it looks very different. A sphere, a torus, a Klein bottle are all examples of such manifolds.

Example 6.3: A sphere can be covered by two co-ordinate patches: one excluding the south pole and the other excluding the north pole. To get the "stereographic co-ordinate" x of a point $P \ne S$ in the first patch, draw a straight line that connects the south pole to P; extend it until it meets the tangent plane to the north pole (see Fig. 6.2). N itself has zero as the co-ordinate. As you move closer to the south pole, the co-ordinate x moves off to infinity. That is why S itself has to be excluded. Interchange N and S to get the other co-ordinate patch.

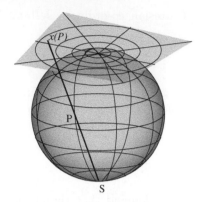

Fig. 6.2 The stereographic co-ordinate on the sphere.

6.3.   Vector fields as derivations

The basic object on a differential manifold is a smooth function; i.e., a function $f : M \to R$ such that its restriction to each path is a smooth function of the co-ordinates. The set of functions in the space M form a commutative algebra over the real number field: we can multiply them by constants (real numbers); and we can multiply two functions to get another:

$$fg(x) = f(x)g(x)$$

A *derivation* of a commutative algebra A is a linear map $V : A \to A$ that satisfies the *Leibnitz rule*

$$V(fg) = (Vf)g + fV(g)$$

We can add two derivations to get another. Also we can multiply them by real numbers to get another derivation. The set of derivations of a commutative algebra form a *module* over A; i.e., the left multiplication fV is also a derivation. In our case of functions in a space, each component of the vector field is multiplied by the scalar f.

In this case, a derivation is the same as a first order differential operator or *vector field*:

$$Vf = V^\mu \partial_\mu f$$

The coefficient of the derivative along each co-ordinate is the component of the vector field in the direction.

The product of two derivations is not in general a derivation: it does not satisfy the Leibnitz rule:

$$V\left(W(fg)\right) = V\left((Wf)g + fW(g)\right)$$

$$= V(W(f))g + 2(Wf)(Vg) + fV(W(g))$$

But the commutator of two derivations, defined as

$$[V, W] = VW - WV$$

is always another derivation:

$$V\left(W(fg)\right) - W\left(V(fg)\right) = V(W(f))g + fV(W(g)) - W(V(f))g - fW(V(g))$$

In terms of components,

$$[V, W]^{\mu} = V^{\nu}\partial_{\nu}W^{\mu} - W^{\nu}\partial_{\nu}V^{\mu}$$

We sum over repeated indices as is the convention in geometry. The commutator of vector fields satisfies the identities

$$[V, W] + [W, V] = 0$$
$$[[U, V], W] + [[V, W], U] + [[W, U], V] = 0$$

Any algebra that satisfies these identities is called a *Lie algebra*. The set of vector fields on a manifold is the basic example. Lie algebras describe infinitesimal transformations. Many of them are of great interest in physics, as they describe symmetries. More on this later.

Example 6.4: The vector field

$$V = \frac{\partial}{\partial x}$$

generates translations along x. Its integral curve is $x(t) = x_0 + t$.

Example 6.5: On the other hand,

$$W = x\frac{\partial}{\partial x}$$

generates scaling. That is, its integral curve is

$$x(u) = e^u x_0$$

We see that these two vector fields have the commutator

$$[V, W] = V$$

Example 6.6: Infinitesimal rotations around the three Cartesian axes on R^3 are described by the vector fields

$$L_x = -y\frac{\partial}{\partial z} + z\frac{\partial}{\partial y}, \quad L_y = -z\frac{\partial}{\partial x} + x\frac{\partial}{\partial z}, \quad L_z = -x\frac{\partial}{\partial y} + y\frac{\partial}{\partial x}$$

They satisfy the commutation relations

$$[L_x, L_y] = L_z, \quad [L_y, L_z] = L_x, \quad [L_z, L_x] = L_y$$

Given two vector fields V, W, we can imagine moving from x_0 along the integral curve of V for a time δ_1 and then along that of W for some time δ_2. Now suppose we reverse the order by first going along the integral curve of W for a time δ_2 and then along that of V for a time δ_2. The difference between the two endpoints is order $\delta_1 \delta_2$, but is not in general zero. It is equal to the commutator:

$$[V^\nu \partial_\nu W^\mu - W^\nu \partial_\nu V^\mu] \, \delta_1 \delta_2$$

Thus, the commutator of vector fields, which we defined above algebraically, has a geometric meaning as the difference between moving along the integral curves in two different ways. Such an interplay of geometry and algebra enriches both fields. Usually geometry helps us imagine things better or relate the mathematics to physical situations. The algebra is more abstract and allows generalizations to new physical situations that were previously unimaginable. For example, the transition from classical to quantum mechanics involves non-commutative algebra. These days we are circling back and constructing a new kind of geometry, *non-commutative geometry*, which applies to quantum systems (Connes, 1994).

6.4. Fixed points

A point at which a vector field vanishes is called a *fixed point*. An orbit that starts there stays there. Thus equilibrium points of dynamical systems are fixed points. It is interesting to ask what happens if we are close to but not at a fixed point. Being differentiable, the components of the vector field can be expanded in a Taylor series. It makes sense to choose a co-ordinate system whose origin is the fixed point. Thus

$$V^\mu(x) = V'^\mu_\nu x^\nu + \mathrm{O}(x^2)$$

The matrix V' is the main actor now. Suppose it can be diagonalized: there is an invertible real matrix S such that $V' = S^{-1} \Lambda S$, and Λ is diagonal. Then we can make a linear change of co-ordinates to bring the equations to the form

$$\dot{x}^\mu = \lambda_\mu x^\mu + \mathrm{O}(x^2)$$

The integral curves are,

$$x^\mu(t) \approx A_\mu e^{\lambda_\mu t}$$

This remains small, and the approximation remains valid, as long as λ_μ and t have opposite signs; or if t is small. If all the eigenvalues are real and negative, we say that the fixed point is stable. Forward time evolution will drive any point near the fixed point even closer to it. If all the eigenvalues are positive, we have an unstable fixed point. Although for small positive times the orbit is nearby, V' does not have enough information to determine the fate of the orbit for large times. Such unstable orbits can for example, be dense (come as close as necessary to every point in phase space).

If some λ are positive and some are negative, there is a subspace of the tangent space at the fixed point containing the stable directions and a complementary one containing the unstable directions. There is a neighborhood of the fixed point which can then be expressed as a product of a stable and an unstable submanifold. But as we move away from the fixed point, these need not be submanifolds at all: they might intersect, or one (or both) of them might be dense. What if some of the eigenvalues vanish? Then the dynamics in those directions is determined by going to the next order in the Taylor expansion. Whether those directions are stable or not depends on the sign of the second derivative of V in those directions.

A physically important case is when the characteristic values are complex.

Example 6.7: Let us return to the damped harmonic oscillator 6.1. Clearly $x = 0$ is a fixed point and

$$V' = \begin{pmatrix} 0 & 1 \\ -\omega^2 & -\gamma \end{pmatrix}$$

The eigenvalues are $\lambda_\pm = -\frac{\gamma}{2} \pm i\sqrt{\omega^2 - \frac{\gamma^2}{4}}$. As long as $\gamma > 0, \omega > \frac{\gamma}{2}$, the orbits are spirals converging to the fixed points:

$$x^1 = e^{-\frac{\gamma}{2}t} A \sin\left[\sqrt{\omega^2 - \frac{\gamma^2}{4}}\, t + \phi\right]$$

Since V' is real, its complex eigenvalues must come in pairs: if λ is an eigenvalue, so will be λ^*. Each such pair is associated to a plane in the tangent space at the fixed point. If $\mathrm{Re}\lambda < 0$, the orbits spiral towards the fixed point as $t \to \infty$. So complex eigenvalues are stable if $\mathrm{Re}\lambda < 0$ and unstable if $\mathrm{Re}\lambda > 0$.

Problem 6.1: What are the orbits of the over-damped harmonic oscillator: Example 6.1 with $\gamma > 0, \omega^2 < \frac{\gamma^2}{4}$? What are the orbits when $\gamma < 0$ but $\omega^2 > \frac{\gamma^2}{4}$? Also, analyze the case where $\gamma > 0, \omega^2 > \frac{\gamma^2}{4}$.

Problem 6.2: Consider the torus $\mathbb{T} = \mathbb{S}^1 \times \mathbb{S}^1$. We can think of it as the set of points on the plane modulo translation by integers. Define the vector field $V = (1, \gamma)$ where γ is a real number. Show that if $\gamma = \frac{m}{n}$ (for co-prime m, n) is a rational number, the orbit is a closed curve that winds m times around the first circle and n times around the second circle. Also, that if γ is irrational, the orbit is dense: every point on the torus is within a distance ϵ of some point on the orbit, for any ϵ.

Problem 6.3: Consider a two-dimensional ODE with a fixed point at which the characteristic polynomial of the Jacobi matrix has unequal roots. Recall that in this case, there is a change of co-ordinates which will bring this matrix to diagonal form. Plot the orbits in each of the following cases:

1. The roots are real and non-zero: cases when they are both positive, one is positive and one negative and both are negative.
2. One root is zero and the cases when the other root is positive or negative.
3. If the roots are complex, they must be complex conjugates of each other. Plot the orbits for the cases when the real part is positive, zero and negative.

Problem 6.4*: Let $\beta(\alpha)\frac{d}{d\alpha}$ be a vector field on the real line with a double zero at the origin:

$$\beta(\alpha) = \beta_2\alpha^2 + \beta_3\alpha^3 + \beta_4(\alpha)\alpha^4$$

$\beta_2 \neq 0, \beta_3$ are constants and $\beta_4(\alpha)$ is a smooth function in some neighborhood of the origin. Show that there is a change of variables

$$x(\alpha) = x_1\alpha + x_3(\alpha)\alpha^3$$

such that the vector field can be brought to the standard form

$$\beta(\alpha)\frac{d}{d\alpha} = [1 + \lambda x]\,x^2\frac{d}{dx}, \quad \lambda = \frac{\beta_3}{\beta_2^2}$$

Here, $x_3(\alpha)$ should be a smooth function in a neighborhood of the origin. We require that $x_1 \neq 0$ so that the change of variables $\alpha \mapsto x$ is invertible in a neighborhood of the origin. It will reverse the orientation if $x_1 < 0$ and not otherwise.

Solution
We get

$$\left[\beta_2\alpha^2 + \beta_3\alpha^3 + \beta_4(\alpha)\alpha^4\right]\frac{dx}{d\alpha} = x^2(\alpha) + \lambda x^3(\alpha)$$

Equating terms of order α^2

$$\beta_2 x_1 = x_1^2 \implies x_1 = \beta_2$$

In order α^3 we get

$$\beta_3 x_1 = \lambda x_1^3$$

Putting these in and simplifying we get a differential equation

$$\left[\beta_2 + \beta_3\alpha + \beta_4(\alpha)\alpha^2\right]x_3'(\alpha) + \beta_2\frac{\beta_4(\alpha) + x_3(\alpha)}{\alpha} +$$

$$\alpha x_3(\alpha)[3\beta_4(\alpha) - x_3(\alpha)] - \frac{\beta_3\alpha^2 x_3(\alpha)^2}{\beta_2^2}\left[3\beta_2 + \alpha^2 x_3(\alpha)\right] = 0$$

Fig. 6.3 The Lorenz system.

The singularity at $\alpha = 0$ cancels out for the initial condition

$$x_3(0) = -\beta_4(0)$$

yielding a solution in some neighborhood of $\alpha = 0$. This problem arises in the study of asymptotically free quantum field theories by G. 't Hooft.

Problem 6.5: The Lorenz system is defined by

$$\dot{x} = \alpha(y - x), \quad \dot{y} = -xz + \beta x - y, \quad \dot{z} = xy - z$$

Write a computer program that solves it numerically and plots the integral curves. Of special interest is to study the case $\beta = 20$ with $\alpha = 3$ for the time range $0 < t < 250$ and initial condition $x(0) = 0 = z(0), y(0) = 1$. The orbit lies on a submanifold called a strange attractor, shaped like the wings of a butterfly (see Fig. 6.3). A small change in time can result in the system switching from one wing to the other.

7
Hamilton's principle

William Rowan Hamilton (1805–1865) was the Astronomer Royal for Ireland. In this capacity, he worked on two important problems of mathematical interest: the motion of celestial bodies and the properties of light needed to design telescopes. Amazingly, he found that the laws of mechanics and those of ray optics were, in the proper mathematical framework, remarkably similar. But ray optics is only an approximation, valid when the wavelength of light is small. He probably wondered in the mid nineteenth century: could mechanics be the short wavelength approximation of some wave mechanics?

The discovery of quantum mechanics brought this remote outpost of theoretical physics into the very center of modern physics.

7.1. Generalized momenta

Recall that to each co-ordinate q^i we can associate a momentum variable,

$$p_i = \frac{\partial L}{\partial \dot{q}^i}$$

p_i is said to be *conjugate* to q^i. It is possible to eliminate the velocities and write the equations of motion in terms of q^i, p_i. In this language, the equations will be a system of first order ODEs. Recall that from the definition of the hamiltonian

$$L = \sum_i p_i \dot{q}^i - H$$

So if we view $H(q, p)$ as a function of position and momentum, we get a formula for the action

$$S = \int \left[\sum_i p_i \dot{q}^i - H(q, p, t) \right] dt$$

Suppose we find the condition for the action to be an extremum, treating q^i, p_i as independent variables:

$$\delta S = \int \sum_i \left[\delta p_i \dot{q}^i + p_i \frac{d}{dt} \delta q^i - \delta p_i \frac{\partial H}{\partial p_i} - \delta q^i \frac{\partial H}{\partial q^i} \right] dt = 0$$

We get the system of ODE

$$\frac{dq^i}{dt} = \frac{\partial H}{\partial p_i}, \quad \frac{dp_i}{dt} = -\frac{\partial H}{\partial q^i}$$

These are called *Hamilton's equations*. They provide an alternative formulation of mechanics.

Example 7.1: A particle moving on the line under a potential has $L = \frac{1}{2}m\dot{q}^2 - V(q)$ and $H = \frac{p^2}{2m} + V(q)$
It follows that

$$\frac{dq}{dt} = \frac{p}{m}, \quad \frac{dp}{dt} = -V'(q)$$

Clearly, these are equivalent to Newton's second law.

Example 7.2: For the simple pendulum

$$L = \frac{1}{2}\dot{\theta}^2 - [1 - \cos\theta], \quad H = \frac{p_\theta^2}{2} + 1 - \cos\theta$$

In terms of the variable $x = \frac{1}{k}\sin\frac{\theta}{2}$

$$L = 2k^2 \left[\frac{\dot{x}^2}{1 - k^2 x^2} - x^2 \right]$$

It follows that

$$p = \frac{\partial L}{\partial \dot{x}} = 4k^2 \frac{\dot{x}}{1 - k^2 x^2}, \quad H = 2k^2 \left[\left(1 - k^2 x^2\right) p^2 + x^2 \right]$$

If we choose the parameter k such that $H = 2k^2$ the relation between p and x becomes

$$p^2 = \frac{1 - x^2}{1 - k^2 x^2}$$

This is another description of the elliptic curve, related rationally to the more standard one:

$$y = \left(1 - k^2 x^2\right)\frac{p}{4k^2}, \quad y^2 = \left(1 - x^2\right)\left(1 - k^2 x^2\right)$$

Example 7.3: A particle moving under a potential in three dimensions has

$$L = \frac{1}{2}m\dot{\mathbf{r}}^2 - V(\mathbf{r})$$

so that

$$H = \frac{\mathbf{p}^2}{2m} + V(\mathbf{r})$$

$$\dot{\mathbf{r}} = \frac{\mathbf{p}}{m}, \quad \dot{\mathbf{p}} = -\nabla V$$

For the Kepler problem

$$V(\mathbf{r}) = -\frac{GMm}{|\mathbf{r}|}$$

7.2. Poisson brackets

Being first order ODE, the solution for Hamilton's equations is determined once the value of (q^i, p_i) is known at one instant. The space M whose co-ordinates are (q^i, p_i) is the *phase space*. Each point of phase space determines a solution of Hamilton's equation, which we call the *orbit* through that point. Hamilton's equations tell us how a given point in phase space evolves under an infinitesimal time translation: they define a vector field in the phase space. By compounding such infinitesimal transformations, we can construct time evolution over finite time intervals: the orbit is the *integral curve* of Hamilton's vector field.

$$V_H = \frac{\partial H}{\partial p_i}\frac{\partial}{\partial q^i} - \frac{\partial H}{\partial q_i}\frac{\partial}{\partial p_i}$$

Since the state of the system is completely specified by a point in the phase space, any physical observable must be a function $f(q, p)$; that is, a function $f : M \to R$ of position and momentum.[1] The hamiltonian itself is an example of an observable; perhaps the most important one.

We can work out an interesting formula for the total time derivative of an observable:

$$\frac{df}{dt} = \sum_i \left[\frac{dq^i}{dt}\frac{\partial f}{\partial q^i} + \frac{dp_i}{dt}\frac{\partial f}{\partial p_i} \right]$$

Using Hamilton's equations this becomes

$$\frac{df}{dt} = \sum_i \left[\frac{\partial H}{\partial p_i}\frac{\partial f}{\partial q^i} - \frac{\partial H}{\partial q^i}\frac{\partial f}{\partial p_i} \right]$$

Given any pair of observables, we define their *Poisson bracket* to be

$$\{g, f\} = \sum_i \left[\frac{\partial g}{\partial p_i}\frac{\partial f}{\partial q^i} - \frac{\partial g}{\partial q^i}\frac{\partial f}{\partial p_i} \right]$$

Thus

$$\frac{df}{dt} = \{H, f\}$$

---

[1] Sometimes we would also allow an explicit dependence in time, but we ignore that possibility for the moment.

A particular example is when f is one of the co-ordinates themselves:

$$\{H, q^i\} = \frac{\partial H}{\partial p_i}, \quad \{H, p_i\} = -\frac{\partial H}{\partial q^i}$$

You may verify the *canonical relations*

$$\{p_i, p_j\} = 0 = \{q^i, q^j\}, \quad \{p_i, q^j\} = \delta_i^j$$

Here $\delta_i^j = \begin{cases} 1 & \text{if } i = j \\ 0 & \text{otherwise} \end{cases}$ is the *Kronecker symbol*.

Exercise 7.1: Show that the Poisson bracket satisfies the conditions for a Lie algebra:

$$\{f, g\} + \{g, f\} = 0, \quad \{\{f, g\}, h\} + \{\{g, h\}, f\} + \{\{h, f\}, g\} = 0$$

and in addition that

$$\{f, gh\} = \{f, g\} h + g \{f, h\}$$

Together these relations define a *Poisson algebra*.

7.3. The star product*

The observables of quantum mechanics can still be thought of as functions in the phase space. But then, the rule for multiplying observables is no longer the obvious one: it is a non-commutative operation. Without explaining how it is derived, we can exhibit the formula for this quantum multiplication law in the case of one degree of freedom:

$$f*g = fg - \frac{i\hbar}{2}\left[\frac{\partial f}{\partial p}\frac{\partial g}{\partial q} - \frac{\partial f}{\partial q}\frac{\partial g}{\partial p}\right] + \frac{1}{2}\left(\frac{i\hbar}{2}\right)^2\left[\frac{\partial^2 g}{\partial p^2}\frac{\partial^2 f}{\partial q^2} - \frac{\partial^2 g}{\partial q^2}\frac{\partial^2 f}{\partial p^2}\right] + \cdots$$

Or,

$$f*g = fg + \sum_{r=1}^{\infty} \frac{1}{r!}\left(-\frac{i\hbar}{2}\right)^r\left[\frac{\partial^{2r} g}{\partial p^r}\frac{\partial^r f}{\partial q^r} - \frac{\partial^r g}{\partial q^r}\frac{\partial^r f}{\partial p^r}\right]$$

This is an associative, but not commutative, multiplication: $(f*g)*h = f*(g*h)$. (It can be proved using a Fourier integral representation.) Note that the zeroth order term is the usual multiplication and that the Poisson bracket is the first correction. In particular, in quantum mechanics we have the Heisenberg commutation relations

$$[p, q] = -i\hbar$$

where the *commutator* is defined as $[p, q] = p*q - q*p$.

The constant functions are multiples of the identity in this $*$-product. We can define the inverse, exponential and trace of an observable under the $*$-product by

$$ f * f^{*-1} = 1, \quad \exp_* f = 1 + \sum_{r=1}^{\infty} \frac{1}{r!} f * f * \cdots (r \text{ times}) f, \quad \mathrm{tr} f = \int f(q,p) \, \frac{dq dp}{(2\pi\hbar)^n} $$

The trace may not exist always: the identity has infinite trace. An eigenvalue of f is a number λ for which $f - \lambda$ does *not* have an inverse. Any quantum mechanics problem solved in the usual Heisenberg or Schrödinger formulations can be solved in this method as well. Which method you use is largely a matter of convenience and taste.

Thus, Poisson algebras are approximations to non-commutative but associative algebras. Is there a non-commutative generalization of geometric ideas such as co-ordinates and vector fields? This is the subject of non-commutative geometry, being actively studied by mathematicians and physicists. This approach to quantization, which connects hamiltonian mechanics to Heisenberg's formulation of quantum mechanics, is called *deformation quantization*. Every formulation of classical mechanics has its counterpart in quantum mechanics; each such bridge between the two theories is a convenient approach to certain problems. Deformation quantization allows us to discover not only non-commutative geometry but also new kinds of symmetries of classical and quantum systems where the rules for combining conserved quantities of isolated systems is non-commutative: *quantum groups*. This explained why certain systems that did not have any obvious symmetry could be solved by clever folks such as Bethe, Yang and Baxter. Once the principle is discovered, it allows solution of even more problems. But now we are entering deep waters.

7.4. Canonical transformation

Suppose we make a change of variables

$$ q^i \mapsto Q^i(q,p), \quad p_i \mapsto P_i(q,p) $$

in the phase space. What happens to the Poisson brackets of a pair of observables under this? Using the chain rule of differentiation

$$ \frac{\partial f}{\partial q^i} = \frac{\partial f}{\partial Q^j} \frac{\partial Q^j}{\partial q^i} + \frac{\partial f}{\partial P_j} \frac{\partial P_j}{\partial q^i} $$

$$ \frac{\partial f}{\partial p_i} = \frac{\partial f}{\partial Q^j} \frac{\partial Q^j}{\partial p_i} + \frac{\partial f}{\partial P_j} \frac{\partial P_j}{\partial p_i} $$

Using this, and some elbow grease, you can show that

$$ \{f,g\} = \left\{Q^i, Q^j\right\} \frac{\partial f}{\partial Q^i} \frac{\partial g}{\partial Q^j} + \{P_i, P_j\} \frac{\partial f}{\partial P_i} \frac{\partial g}{\partial P_j} + \left\{P_i, Q^j\right\} \left\{ \frac{\partial f}{\partial P_i} \frac{\partial g}{\partial Q^j} - \frac{\partial f}{\partial Q^j} \frac{\partial g}{\partial P_i} \right\} $$

So if the new variables happen to satisfy the canonical relations as well:

$$ \{P_i, P_j\} = 0 = \left\{Q^i, Q^j\right\}, \quad \left\{P_i, Q^j\right\} = \delta_i^j $$

the Poisson brackets are still given by a similar expression:

$$\{f, g\} = \sum_i \left[\frac{\partial f}{\partial P_i} \frac{\partial g}{\partial Q^i} - \frac{\partial f}{\partial Q^i} \frac{\partial g}{\partial P_i} \right]$$

Such transformations are called *canonical transformations*; they are quite useful in mechanics because they preserve the mathematical structure of mechanics. For example, Hamilton's equations remain true after a canonical transformation:

$$\frac{dQ^i}{dt} = \frac{\partial H}{\partial P_i}$$

$$\frac{dP_i}{dt} = -\frac{\partial H}{\partial Q_i}$$

Example 7.4: *The case of one degree of freedom.* The interchange of position and momentum variables is an example of a canonical transformation:

$$P = -q, \quad Q = p$$

Notice the sign.

Another example is the scaling

$$Q = \lambda q, \quad P = \frac{1}{\lambda} p$$

Notice the inverse powers. More generally, the condition for a transformation $(q, p) \mapsto (Q, P)$ to be canonical is that the area element $dq\,dp$ be transformed to $dQ\,dP$. This is because, in the case of one degree of freedom, the Poisson bracket happens to be the Jacobian determinant:

$$\{P, Q\} \equiv \frac{\partial P}{\partial p} \frac{\partial Q}{\partial q} - \frac{\partial Q}{\partial p} \frac{\partial P}{\partial q} = \det \begin{bmatrix} \dfrac{\partial Q}{\partial q} & \dfrac{\partial Q}{\partial p} \\ \dfrac{\partial P}{\partial q} & \dfrac{\partial P}{\partial p} \end{bmatrix}$$

For more degrees of freedom, it is still true that the volume element in phase space in invariant, $\prod_i dq^i\,dp_i = \prod_i dQ^i\,dP_i$, under canonical transformations, a result known as Liouville's theorem. But the invariance of the phase space volume no longer guarantees that a transformation is canonical: the conditions for that are stronger.

7.5.   Infinitesimal canonical transformations

The composition of two canonical transformations is also a canonical transformation. Sometimes we can break up a canonical transformation as the composition of infinitesimal transformations. For example, when $\lambda > 0$, the following is a canonical transformation.

$$Q = \lambda q, \quad P = \lambda^{-1} p$$

A scaling through λ_1 followed by another through λ_2 is equal to one by $\lambda_1\lambda_2$. Conversely, we can think of a scaling through $\lambda = e^\theta$ as made of up of a large number n of scalings, each through a small value $\frac{\theta}{n}$. For an infinitesimally small $\Delta\theta$,

$$\Delta Q = q\Delta\theta, \quad \Delta P = -p\Delta\theta$$

These infinitesimal changes of co-ordinates define a vector field

$$V = q\frac{\partial}{\partial q} - p\frac{\partial}{\partial p}$$

That is, the effect of an infinitesimal rotation on an arbitrary observable is

$$Vf = q\frac{\partial f}{\partial q} - p\frac{\partial f}{\partial p}$$

Now, note that this can be written as

$$Vf = \{pq, f\}$$

This is a particular case of a more general fact: every infinitesimal canonical transformation can be thought of as the Poisson bracket with some function, called its *generating function*.

Remark 7.1: Infinitesimal canonical transformations are called hamiltonian vector fields by mathematicians, the generating function being called the hamiltonian. You have to be careful to remember that in physics, hamiltonian means the generator of a particular such transformation, time evolution. Mathematics and physics are two disciplines divided by a common language.

Let us write an infinitesimal canonical transformation in terms of its components

$$V = V^i\frac{\partial}{\partial q^i} + V_i\frac{\partial}{\partial p_i}$$

The position of the indices is chosen such that V^i is the infinitesimal change in q^i and V_i the change in p_i.

$$q^i \mapsto Q^i = q^i + V^i\Delta\theta, \quad p_i \mapsto P_i = p_i + V_i\Delta\theta$$

for some infinitesimal parameter $\Delta\theta$. Let us calculate to first order in $\Delta\theta$:

$$\{Q^i, Q^j\} = (\{V^i, q^j\} + \{q^i, V^j\})\,\Delta\theta$$
$$\{P_i, P_j\} = (\{V_i, p_j\} + \{p_i, V_j\})\,\Delta\theta$$
$$\{P_i, Q^j\} = \delta_i^j + (\{V_i, q^j\} + \{p_i, V^j\})\,\Delta\theta$$

So the conditions for V to be an infinitesimal canonical transformation are

$$\{V^i, q^j\} + \{q^i, V^j\} = 0$$
$$\{V_i, p_j\} + \{p_i, V_j\} = 0$$
$$\{V_i, q^j\} + \{p_i, V^j\} = 0$$

In terms of partial derivatives

$$\frac{\partial V^i}{\partial p_j} - \frac{\partial V^j}{\partial p_i} = 0$$

$$\frac{\partial V_i}{\partial q^j} - \frac{\partial V_j}{\partial q^j} = 0$$

$$\frac{\partial V_i}{\partial p_j} + \frac{\partial V^j}{\partial q^i} = 0 \qquad (7.1)$$

The above conditions are satisfied if[2]

$$V^i = \frac{\partial G}{\partial p_i}, \quad V_i = -\frac{\partial G}{\partial q^i} \qquad (7.2)$$

for some function G. The proof is a straightforward computation of second partial derivatives:

$$\frac{\partial V^i}{\partial p_j} - \frac{\partial V^j}{\partial p_i} = \frac{\partial^2 G}{\partial p_i \partial p_j} - \frac{\partial^2 G}{\partial p_i \partial p_j} = 0$$

etc.

Conversely, if (7.1) implies (7.2), we can produce the required function f as a line integral from the origin to the point (q, p) along some curve:

$$G(q,p) = \int_{(0,0)}^{(q,p)} \left[V_i \frac{dq^i}{ds} - V^i \frac{dp_i}{ds} \right] ds$$

In general such integrals will depend on the path taken, not just the endpoint. But the conditions (7.1) are exactly what is needed to ensure independence on the path.[3]

Exercise 7.2: Prove this by varying the path infinitesimally.

Thus an infinitesimal canonical transformation is the same as the Poisson bracket with some function, called its generator. By composing such infinitesimal transformations, we get a curve on the phase space:

$$\frac{dq^i}{d\theta} = \{G, q^i\}, \quad \frac{dp_i}{d\theta} = \{G, p_i\}$$

---

[2] There is an analogy with the condition that the curl of a vector field is zero; such a vector field would be the gradient of a scalar.

[3] We are assuming that any two curves connecting the origin to (q,p) can be deformed continuously into each other. In topology, the result we are using is called the Poincaré lemma.

Now we see that Hamilton's equations are just a special case of this. Time evolution is a canonical transformation too, whose generator is the hamiltonian. Every observable (i.e., function in phase space) generates its own canonical transformation.

Example 7.5: A momentum variable generates translations in its conjugate position variable.

Example 7.6: The generator of rotations is angular momentum along the axis of rotation. For a rotation around the z-axis

$$x \mapsto \cos\theta x - \sin\theta y, \quad y \mapsto \sin\theta x + \cos\theta y$$
$$p_x \mapsto \cos\theta p_x + \sin\theta p_y, \quad p_y \mapsto -\sin\theta p_x + \cos\theta p_y$$

So we have

$$\frac{dx}{d\theta} = -y, \quad \frac{dy}{d\theta} = x$$
$$\frac{dp_x}{d\theta} = p_y, \quad \frac{dp_y}{d\theta} = -p_x$$

The generator is

$$L_z = xp_y - yp_x$$

7.6. Symmetries and conservation laws

Suppose that the hamiltonian is independent of a certain co-ordinate q^i; then the corresponding momentum is conserved.

$$\frac{\partial H}{\partial q^i} = 0 \implies \frac{dp_i}{dt} = 0$$

This is the beginning of a much deeper theorem of Noether that asserts that every continuous symmetry implies a conservation law. A symmetry is any canonical transformation of the variables $(q^i, p_i) \mapsto (Q^i, P_i)$ that leaves the hamiltonian unchanged:

$$\{P_i, P_j\} = 0 = \{Q^i, Q^j\}, \quad \{P_i, Q^j\} = \delta_i^j$$
$$H(Q(q,p), P(q,p)) = H(q,p)$$

A continuous symmetry is one that can be built up as a composition of infinitesimal transformations. We saw that every such canonical transformation is generated by some observable G. The change of any other observable f under this canonical transformation is given by

$$\{G, f\}$$

In particular the condition that the hamiltonian be unchanged is

$$\{G, H\} = 0$$

But we saw earlier that the change of G under a time evolution is

$$\frac{dG}{dt} = \{H, G\}$$

So, the invariance of H under the canonical transformation generated by G is equivalent to the condition that G is conserved under time evolution.

$$\frac{dG}{dt} = 0 \iff \{G, H\} = 0$$

Example 7.7: Let us return to the Kepler problem, $H = \frac{\mathbf{p}^2}{2m} + V(\mathbf{r})$, where $V(\mathbf{r})$ is a function only of the distance $|\mathbf{r}|$. The components L_x, L_y, L_z of angular momentum

$$\mathbf{L} = \mathbf{r} \times \mathbf{p}$$

generate rotations around the axes x, y, z respectively. Since the hamiltonian is invariant under rotations

$$\{\mathbf{L}, H\} = 0$$

Thus the three components of angular momentum are conserved:

$$\frac{d\mathbf{L}}{dt} = 0$$

This fact can also be verified directly as we did before.

7.7. Generating function

Suppose that $(p_i, q^i) \mapsto (P_i, Q^i)$ is a canonical transformation; that is, the functions $p_i(P, Q), q^i(P, q)$ satisfy the differential equations

$$\sum_k \left(\frac{\partial p_i}{\partial P_k} \frac{\partial p_j}{\partial Q^k} - \frac{\partial p_j}{\partial P_k} \frac{\partial p_i}{\partial Q^k} \right) = 0 = \sum_k \left(\frac{\partial q^i}{\partial P_k} \frac{\partial q^j}{\partial Q^k} - \frac{\partial q^j}{\partial P_k} \frac{\partial q^i}{\partial Q^k} \right)$$

$$\sum_k \left(\frac{\partial p_i}{\partial P_k} \frac{\partial q^j}{\partial Q^k} - \frac{\partial q^j}{\partial P_k} \frac{\partial p_i}{\partial Q^k} \right) = \delta_i^j$$

Clearly this is an over-determined system: there are more equations $(2n^2 - n)$ than unknowns $(2n)$. The reason there are solutions is that not all of them are independent. In fact, there are many solutions of this system, each determined by a function of $2n$ variables.

Suppose we pick a function $F(P,q)$ and solve the following system to determine p, q as functions of P, Q.

$$p_i = \frac{\partial F(P,q)}{\partial q^i}, \quad Q^i = \frac{\partial F(P,q)}{\partial P_i}$$

Then $p_i(P,Q), q^i(P,q)$ determined this way will automatically satisfy the canonical Poisson brackets. In order to be able to solve the above equation locally, the Hessian matrix $\frac{\partial^2 F(P,q)}{\partial P_i \partial q^j}$ must be invertible.

Example 7.8: The identity transformation is generated by $F(P,q) = P_i q^i$.

Example 7.9: If $F(P,q) = \lambda P_i q^i$, we get $p_i = \lambda P_i, q^i = \lambda^{-1} Q^i$. That is, it generates scaling.

Example 7.10: A particular case is when $q(P,Q)$ only depends on Q and not on P; that is a change of co-ordinates of the configuration space. Then $F(P,q) = P_i Q^i(q)$ where $Q(q)$ is the inverse function of $q(Q)$. Then $p_i(P,Q)$ is determined by eliminating q in favor of Q on the RHS of

$$p_i = P_j \frac{\partial Q^j(q)}{\partial q^i}$$

For a deeper study of the group theoretical meaning of canonical transformations, see Sudarshan and Mukunda (1974).

Problem 7.3: Show that the hamiltonian of the Kepler problem in spherical polar co-ordinates is

$$H = \frac{p_r^2}{2m} + \frac{L^2}{2mr^2} - \frac{k}{r}, \quad L^2 = p_\theta^2 + \frac{p_\phi^2}{\sin^2 \theta}$$

Show that L is the magnitude of angular momentum and that $p_\phi = L_z$ is one of its components. Thus, show that $\{L^2, L_z\} = 0 = \{H, L^2\}$.

Problem 7.4: Not all symmetries of equations of motion lead to conservation laws. Show that if $\mathbf{r}(t)$ is a solution to the equations of the Kepler problem, and so is $\lambda^{\frac{2}{3}} \mathbf{r}(\lambda t)$ for any $\lambda > 0$. How should $\mathbf{p}(t)$ change under this transformation so that Hamilton's equations are preserved? Are the canonical commutation relations preserved? Why can't you construct a conservation law using this symmetry?

Problem 7.5: Let $h_t = \exp_*(-tH)$ for some function H. Show that it satisfies the differential equation

$$\frac{\partial}{\partial t} h_t(q,p) = -H * h_t(q,p), \quad h_0(q,p) = 1$$

Solve this differential equation in the case of the harmonic oscillator $H = \frac{1}{2} p^2 + \frac{1}{2} \omega^2 q^2$. (Use a Gaussian ansatz $h_t(q,p) = e^{\alpha(t)p^2 + \beta(t)q^2 + \gamma(t)}$ in terms of the *ordinary* exponential and determine the functions α, β, γ.) Show that the trace $\operatorname{tr} h_t = \sum_{n=0} e^{-\hbar\omega(n+\frac{1}{2})}$ and hence that $\operatorname{tr}(H - \lambda)^{*-1} = \sum_{n=0}^{\infty} \frac{1}{\hbar\omega(n+\frac{1}{2})-\lambda}$. This is one way to determine the spectrum of the quantum harmonic oscillator.

Problem 7.6: Show that the generating function of the canonical transformation from Cartesian (x_1, x_2) to polar co-ordinates (r, ϕ) in the plane is

$$p_r \sqrt{x_1^2 + x_2^2} + p_\phi \arctan \frac{x_2}{x_1}$$

Problem 7.7: Show that the rotation through some angle t in the phase space,

$$P = \cos t \; p + \sin t \; q, \quad Q = -\sin t \; p + \cos t \; q$$

is a canonical transformation. What is its generating function $F(P, q)$?

8

Geodesics

A basic problem in geometry is to find the curve of shortest length that passes through two given points. Such curves are called *geodesics*. On the plane, this is a straight line. But if we look at some other surface, such as the sphere, the answer is more intricate. Gauss developed a general theory for geodesics on surfaces, which Riemann generalized to higher dimensions. With the discovery of relativity it became clear that space and time are to be treated on the same footing. Einstein discovered that the Riemannian geometry of space-time provides a relativistic theory of gravitation. Thus, Riemannian geometry is essential to understand the motion of particles in a gravitational field.

The theory of geodesics can be thought of a hamiltonian system, and ideas from mechanics are useful to understand properties of geodesics. In another direction, it turns out that the motion of even non-relativistic particles of a given energy in a potential can be understood as geodesics of a certain metric (Maupertuis metric). Thus, no study of mechanics is complete without a theory of geodesics.

8.1.   The metric

Let x^i for $i = 1, \cdots, n$ be the co-ordinates on some space. In Riemannian geometry, the distance ds between two nearby points x^i and $x^i + dx^i$ is postulated to be a quadratic form[1]

$$ds^2 = g_{ij}(x)dx^i dx^{\nu j}$$

For Cartesian co-ordinates in Euclidean space, g_{ij} are constants,

$$ds^2 = \sum_i \left[dx^i \right]^2, \quad g_{ij} = \delta_{ij}$$

The matrix $g_{ij}(x)$ is called the *metric*. The metric must be a symmetric matrix with an inverse. The inverse is denoted by g^{ij}, with superscripts. Thus

$$g^{ij} g_{jk} = \delta^i_k$$

---

[1]It is a convention in geometry to place the indices on co-ordinates above, as superscripts. Repeated indices are summed over. Thus $g_{ij}(x)dx^i dx^j$ stands for $\sum_{ij} g_{ij}(x)dx^i dx^{\nu j}$. For this to make sense, you have to make sure that no index occurs more than twice in any factor.

Although Riemann only allowed for positive metrics, we now know that the metric of space-time is not positive: along the time-like directions, ds^2 is positive and along space-like directions it is negative. We still require that the metric g_{ij} be a symmetric invertible matrix.

8.2.   The variational principle

A curve is given parametrically by a set of functions $x^i(\tau)$ of some real parameter. The length of this curve will be

$$l[x] = \int \sqrt{g_{ij}(x)\frac{dx^i}{d\tau}\frac{dx^{\nu j}}{d\tau}}d\tau$$

This is the quantity to be minimized, if we are to get an equation for geodesics. But it is simpler (and turns out to be equivalent) to minimize instead the related quantity (the *action* of a curve)

$$S = \frac{1}{2}\int g_{ij}(x)\frac{dx^i}{d\tau}\frac{dx^j}{d\tau}d\tau$$

> **Remark 8.1:** Some mathematicians, making a confused analogy with mechanics, call S the 'energy' of the curve instead of its action.

8.2.1   Curves minimizing the action and the length are the same

If we look at a finite sum instead of an integral, $\frac{1}{2}\sum x_a^2$ and $\sum|x_a|$ are minimized by the same choice of x_a. But x_a^2 is a much nicer quantity than $|x_a|$: for example, it is differentiable.

Similarly, S is a more convenient quantity to differentiate, with the same extremum as the length. This can be proved using a trick using Lagrange multipliers. First of all, we note that the length can be thought of as the minimum of

$$S_1 = \frac{1}{2}\left[\int g_{ij}\frac{dx^i}{d\tau}\frac{dx^j}{d\tau}\lambda^{-1}d\tau + \int \lambda d\tau\right]$$

over all non-zero functions λ. Minimizing gives $\lambda^{-2}|\dot{x}|^2 = 1 \implies \lambda = |\dot{x}|$. At this minimum $S_1[x] = l[x]$. Now S_1 is invariant under changes of parameters

$$\tau \to \tau'(\tau), \lambda' = \lambda\frac{d\tau}{d\tau'}$$

Choosing this parameter to be the arc length, S_1 reduces to the action. Thus they describe equivalent variational problems. Moreover, at the minimum S, S_1, l all agree.

8.2.2   The geodesic equation

Straightforward application of the Euler–Lagrange equation

$$\frac{d}{d\tau}\frac{\partial L}{\partial \dot{x}^i} - \frac{\partial L}{\partial x^i} = 0$$

with Lagrangian

$$L = \frac{1}{2}g_{jk}\dot{x}^j\dot{x}^k$$

leads to

$$\frac{\partial L}{\partial \dot{x}^i} = g_{ij}\dot{x}^j$$

and the geodesic equation

$$\frac{d}{d\tau}\left[g_{ij}\frac{dx^j}{d\tau}\right] - \frac{1}{2}\partial_i g_{jk}\frac{dx^j}{d\tau}\frac{dx^k}{d\tau} = 0$$

An equivalent formulation is

$$\frac{d^2 x^i}{d\tau^2} + \Gamma^i_{jk}\frac{dx^j}{d\tau}\frac{dx^k}{d\tau} = 0, \quad \Gamma^i_{jk} = \frac{1}{2}g^{il}\left[\partial_j g_{kl} + \partial_k g_{jl} - \partial_l g_{jk}\right]$$

The Γ^i_{jk} are called Christoffel symbols. Calculating them for some given metric is one of the joys of Riemannian geometry; an even greater joy is to get someone else to do the calculation for you.

> **Proposition 8.2:** Given an initial point P and a vector V at that point, there is a geodesic that starts at P with V as its tangent.
> This just follows from standard theorems about the local existence of solutions of ODEs. The behavior for large τ can be complicated: geodesics are chaotic except for metrics with a high degree of symmetry.
> The following are more advanced points that you will understand only during a second reading, or after you have already learned some Riemannian geometry.

> **Proposition 8.3:** On a connected manifold, any pair of points are connected by at least one geodesic.
> Connected means that there is a continuous curve connecting any pair of points (to define these ideas precisely we will need first a definition of a manifold, which we will postpone for the moment). Typically there are several geodesics connecting a pair of points, for example on the sphere there are at least two for every (unequal) pair of points: one direct route and that goes around the world.

Proposition 8.4: The shortest length of all the geodesics connecting a pair of points is the distance between them.

It is a deep result that such a minimizing geodesic exists. Most geodesics are extrema, not always minima.

Gauss and Riemann realized that only experiments can determine whether space is Euclidean. They even commissioned an experiment to look for departures from Euclidean geometry; and found none. The correct idea turned out to be to include time as well.

8.3.   The sphere

The geometry of the sphere was studied by the ancients. There were two spheres of interest to astronomers: the surface of the Earth and the celestial sphere, upon which we see the stars. Eratosthenes (3rd century BC) is said to have invented the use of the latitude and longitude as co-ordinates on the sphere. The (6th century AD) Sanskrit treatise *Aryabhatiya*, uses this co-ordinate system for the sphere as well (with the city of Ujjaini on the prime meridian) in solving several problems of spherical geometry. Predicting sunrise and sunset times, eclipses, calculating time based on the length of the shadow of a rod, making tables of positions of stars, are all intricate geometric problems.

The metric of a sphere S^2 in polar co-ordinates is

$$ds^2 = d\theta^2 + \sin^2\theta d\phi^2$$

We just have to hold r constant in the expression for distance in R^3 in polar co-ordinates. The sphere was the first example of a curved space. There are no straight lines on a sphere: any straight line of R^3 starting at a point in S^2 will leave it. There are other subspaces of R^3 such as the cylinder or the cone which contain some straight lines. The question arises: what is the shortest line that connects two points on the sphere? Such questions were of much interest to map makers of the nineteenth century, an era when the whole globe was being explored. In the mid nineteenth century Gauss took up the study of the geometry of distances on curved surfaces, metrics which were later generalized by Riemann to higher dimensions.

The action of a curve on the sphere is

$$S = \frac{1}{2}\int \left[\dot{\theta}^2 + \sin^2\theta\dot{\phi}^2\right] d\tau$$

The Euler–Lagrange equations of this variational principle give

$$\delta S = \int \left[\dot{\theta}\delta\dot{\theta} + \sin\theta\cos\theta\dot{\phi}^2\delta\theta + \sin^2\theta\dot{\phi}\delta\dot{\phi}\right] d\tau$$

$$-\ddot{\theta} + \sin\theta\cos\theta\dot{\phi}^2 = 0$$

$$\frac{d}{d\tau}\left[\sin^2\theta\dot{\phi}\right] = 0$$

The key to solving any system of ODEs is to identify conserved quantities. The obvious conserved quantity is

$$p_\phi = \sin^2\theta\,\dot\phi$$

The solution is simplest when $p_\phi = 0$. For these geodesics, ϕ is also a constant. Then, θ is a linear function of τ. These are the meridians of constant longitude ϕ.

Geometrically, the longitudes are the intersection of a plane passing through the center and the two poles with the sphere. More generally, the intersection of a plane that passes through the center with the sphere is called a *great circle*. The longitudes are examples of great circles. The only latitude that is a great circle is the equator.

Proposition 8.5: Any pair of distinct points on the sphere lie on a great circle. A geodesic connecting them is an arc of this great circle.

There is a plane that contains the two points and the center. Its intersection with the sphere gives the great circle containing the two points. We can always choose one of the points as the origin of our co-ordinate system. By a rotation, we can bring the other point to lie along a longitude. This reduces the problem of solving the geodesic equation to the case we already solved above, without solving the general case.

If you are uncomfortable with this slick use of co-ordinate systems, you can solve the general geodesic equation for $p_\phi \neq 0$.

$$-\ddot\theta + \frac{\cos\theta\, p_\phi^2}{\sin^3\theta} = 0$$

Multiply by $\dot\theta$ and integrate once to get

$$\frac{1}{2}\dot\theta^2 + \frac{p_\phi^2}{2\sin^2\theta} = H$$

another constant of motion. Use

$$\dot\theta \equiv \frac{d\theta}{d\tau} = \dot\phi\frac{d\theta}{d\phi} = \frac{p_\phi}{\sin^2\theta}\frac{d\theta}{d\phi}$$

to get

$$\frac{p_\phi^2}{2\sin^4\theta}\left[\frac{d\theta}{d\phi}\right]^2 + \frac{p_\phi^2}{2\sin^2\theta} = H$$

Or

$$\frac{d\phi}{d\theta} = \frac{p_\phi}{\sqrt{2H}}\frac{1}{\sin\theta\sqrt{\sin^2\theta - \frac{p_\phi^2}{2H}}}$$

which can be solved in terms of trig functions.

8.3.1   The sphere can also be identified with the complex plane, with the point at infinity added

Identify the complex plane with the tangent plane to the sphere at the south plane. Given a point on the sphere, we can draw a straight line in R^3 that connects the north pole to that line: continuing that line, we get a point on the complex plane. This is the co-ordinate of the point. Thus the south pole is at the origin and the north point corresponds to infinity.

The metric of S^2 is

$$ds^2 = 4\frac{d\bar{z}dz}{(1+\bar{z}z)^2}, \quad z = \tan\frac{\theta}{2}e^{i\phi}$$

The isometries of the sphere are fractional linear transformations by $SU(2)$

$$z \mapsto \frac{az+b}{cz+d}, \quad \begin{pmatrix} a & b \\ c & d \end{pmatrix}\begin{pmatrix} \bar{a} & \bar{c} \\ \bar{b} & \bar{d} \end{pmatrix} = \begin{pmatrix} 1 & 0 \\ 0 & 1 \end{pmatrix}$$

Exercise 8.1: Verify by direct calculations that these leave the metric unchanged. This is one way of seeing that $SU(2)/\{1,-1\}$ is the group of rotations.

8.4.   Hyperbolic space

The metric of hyperbolic geometry is

$$ds^2 = d\theta^2 + \sinh^2\theta d\phi^2$$

It describes a space of negative curvature. What this means is that two geodesics that start at the same point in slightly different directions will move apart at a rate faster than in Euclidean space. On a sphere, they move apart slower than in Euclidean space so it has positive curvature. Just as the sphere is the set of points at a unit distance from the center,

Proposition 8.6: The hyperboloid is the set of unit time-like vectors in Minkowski geometry $R^{1.2}$. There is the co-ordinate system analogous to the spherical polar co-ordinate system valid in the time-like interior of the light cone:

$$(x^0)^2 - (x^1)^2 - (x^2)^2 = \tau, \quad x^0 = \tau\cosh\theta,\, x^1 = \tau\sinh\theta\cos\phi,\, x^2 = \tau\sinh\theta\sin\phi$$

The Minkowski metric becomes

$$ds^2 = d\tau^2 - \tau^2\left[d\theta^2 + \sinh^2\theta d\phi^2\right]$$

Thus the metric induced on the unit hyperboloid

$$(x^0)^2 - (x^1)^2 - (x^2)^2 \equiv \tau = 1,$$

is the one above.

By a change of variables,

Proposition 8.7: The hyperboloid can also be thought of as the upper half plane with the metric

$$ds^2 = \frac{dx^2 + dy^2}{y^2}, \quad y > 0$$

Proposition 8.8: The isometries are fractional linear transformations with real parameters a, b, c, d:

$$z \mapsto \frac{az + b}{cz + d}, \quad ad - bc = 1$$

Exercise 8.2: Verify that these are symmetries of the metric.

Proposition 8.9: The geodesics are circles orthogonal to the real line.

If two points have the same value of x, the geodesic is just the line parallel to the imaginary axis that contains them. Using the isometry above we can bring any pair of points to this configuration. It is also possible to solve the geodesic equations to see this fact.

8.5.　Hamiltonian formulation of geodesics

The analogy with mechanics is clear in the variational principle of geometry. The Lagrangian

$$L = \frac{1}{2} g_{ij} \frac{dx^i}{d\tau} \frac{dx^j}{d\tau}$$

leads to the "momenta"

$$p_i = g_{ij} \frac{dx^j}{d\tau}$$

The hamiltonian is

$$H = p_i \frac{dx^i}{d\tau} - L$$
$$= \frac{1}{2} g^{ij} p_i p_j$$

Thus H has the physical meaning of half the square of the mass of the particle. It follows that the geodesic equations can be written as

$$p_i = g_{ij} \frac{dx^j}{d\tau}, \quad \frac{d}{d\tau} p_i = \frac{1}{2} [\partial_i g^{jk}] p_j p_k$$

It is obvious from mechanics that if the metric happens to be independent of a coordinate, its conjugate momentum is conserved. This can be used to solve equations for a geodesic on spaces like the sphere which have such symmetries.

A better point of view is to use the Hamilton–Jacobi equation, a first order partial differential equation (PDE). When this equation is separable, the geodesics can be determined explicitly.

8.6. Geodesic formulation of Newtonian mechanics*

In the other direction, there is a way to think of the motion of a particle in a potential with a fixed energy as geodesics. Suppose we have a particle (or collection of particles) whose kinetic energy is given by a quadratic function of co-ordinates

$$T = \frac{1}{2} m_{ij}(q) \frac{dq^i}{dt} \frac{dq^j}{dt}$$

For example, in the three body problem of celestial mechanics, i, j takes values 1 through 9: the first three for the position of the first particle and so on. Then the metric is a constant matrix whose diagonal entries are the masses:

$$m_{ij} = \begin{bmatrix} m_1 1_3 & 0 & 0 \\ 0 & m_2 1_3 & 0 \\ 0 & 0 & m_3 1_3 \end{bmatrix}$$

The first 3×3 block gives the kinetic energy of the first particle, the next that of the second particle and so on.

If the potential energy is $V(x)$ we have the condition for the conservation of energy

$$\frac{1}{2} m_{ij}(q) \frac{dq^i}{dt} \frac{dq^j}{dt} + V(q) = E$$

If we only consider paths of a given energy E, Hamilton's principle takes the form of minimizing

$$S = \int_{t_1}^{t_2} p_i \frac{dq^i}{dt} dt$$

since $\int H dt = E[t_2 - t_1]$ is constant. Solving for p_i in terms of $\dot{q}^i$ this becomes

$$S = \int [E - V(q)] m_{ij}(q) \frac{dq^i}{ds} \frac{dq^j}{ds} ds$$

where the parameter ds is defined by

$$\frac{ds}{dt} = [E - V(q)]$$

This can be thought of as the variational principle for geodesics of the metric

$$g_{ij} = 2[E - V(q)] m_{ij}(q) dq^i dq^j$$

Of course, this only makes sense in the region of space with $E > V(q)$: that is the only part that is accessible to a particle of total energy E. This version of the variational principle is older than Hamilton's and is due to Euler who was building on ideas of Fermat and Maupertuis in ray optics. Nowadays it is known as the Maupertuis principle.

8.6.1   Keplerian orbits as geodesics

Consider the planar Kepler problem with hamiltonian

$$H = \frac{p_r^2}{2} + \frac{p_\phi^2}{2r^2} - \frac{k}{r}$$

The orbits of this system can be thought of as geodesic of the metric

$$ds^2 = 2 \left[E + \frac{k}{r}\right] \left[dr^2 + r^2 d\phi^2\right]$$

There is no singularity in this metric at the collision point $r = 0$: it can be removed ("regularized") by transforming to the co-ordinates ρ, χ:

$$r = \rho^2, \quad \theta = 2\chi, \implies ds^2 = ds^2 = 8 \left[E\rho^2 + k\right] \left[d\rho^2 + \rho^2 d\chi^2\right]$$

This is just what we would have found for the harmonic oscillator (for $E < 0$): the Kepler problem can be transformed by a change of variables to the harmonic oscillator.

When $E = 0$ (parabolic orbits) this is just the flat metric on the plane: parabolas are mapped to straight lines by the above change of variables. For $E > 0$ (hyperbolic orbits) we get a metric of negative curvature and for $E < 0$ (elliptic orbits) one of positive curvature. These curvatures are not constant, however.

8.7. Geodesics in general relativity*

By far the most important application of Riemannian geometry to physics is general relativity (GR), Einstein's theory of gravitation. The gravitational field is described by the metric tensor of space-time.

The path of a particle is given by the geodesics of this metric. Of special importance is the metric of a spherically symmetric mass distribution, called the Schwarschild metric.

$$ds^2 = \left(1 - \frac{r_s}{r}\right) dt^2 - \frac{dr^2}{1 - \frac{r_s}{r}} - r^2 \left(d\theta^2 + \sin^2 \theta d\phi^2\right)$$

The parameter r_s is proportional to the mass of the source of the gravitational field. For the Sun it is about 1 km. To solve any mechanical problem we must exploit conservation laws. Often symmetries provide clues to these conservation laws.

A time-like geodesic satisfies

$$\left(1 - \frac{r_s}{r}\right) \dot{t}^2 - \frac{\dot{r}^2}{1 - \frac{r_s}{r}} - r^2 \left(\dot{\theta}^2 + \sin^2 \theta \dot{\phi}^2\right) = H$$

Here the dot denotes derivatives w.r.t. τ. The constant H has to be positive; it can be chosen to be one by a choice of units of τ.

Proposition 8.10: Translations in t and rotations are symmetries of the Schwarschild metric. The angular dependence is the same as for the Minkowski metric. The invariance under translations in t is obvious.

Proposition 8.11: Thus the energy and angular momentum of a particle moving in this gravitational field are conserved. The translation in t gives the conservation of energy per unit mass

$$E = p_t = \left(1 - \frac{r_s}{r}\right)\dot{t}$$

We can choose co-ordinates such that the geodesic lies in the plane $\theta = \frac{\pi}{2}$. By looking at the second component of the geodesic equation

$$\frac{d}{d\tau}\left[r^2\frac{d\theta}{d\tau}\right] = r^2\sin\theta\cos\theta\left[\frac{d\phi}{d\tau}\right]^2$$

We can rotate the co-ordinate system so that any plane passing through the center corresponds to $\theta = \frac{\pi}{2}$. The conservation of angular momentum, which is a three-vector, implies also that the orbit lies in the plane normal to it. We are simply choosing the z-axis to point along the angular momentum. Thus

$$\left(1 - \frac{r_s}{r}\right)\dot{t}^2 - \frac{\dot{r}^2}{1 - \frac{r_s}{r}} - r^2\dot{\phi}^2 = H$$

Rotations in ϕ lead to the conservation of the third component of angular momentum per unit mass

$$L = -p_\phi = r^2\dot{\phi}.$$

This is an analog of Kepler's law of areas. To determine the shape of the orbit we must determine r as a function of ϕ.

In the Newtonian limit these are conic sections: ellipse, parabola or hyperbola. Let $u = \frac{r_s}{r}$. Then

$$\dot{r} = r'\dot{\phi} = \frac{r'}{r^2}L = -lu'$$

Here prime denotes derivative w.r.t. ϕ. Also $l = \frac{L}{r_s}$. So,

$$\frac{E^2}{1-u} - \frac{l^2u'^2}{1-u} - l^2u^2 = H$$

We get an ODE for the orbit

$$l^2u'^2 = E^2 + H(u-1) - l^2u^2 + l^2u^3$$

This is the Weierstrass equation, solved by the elliptic integral. Since we are interested in the case where the last term (which is the GR correction) is small, a different strategy is more convenient. Differentiate the equation to eliminate the constants:

$$u'' + u = \frac{H}{2l^2} + \frac{3}{2}u^2$$

Proposition 8.12: In the Newtonian approximation the orbit is periodic. The Newtonian approximation is

$$u_0'' + u_0 = \frac{H}{2l^2} \implies$$

$$u_0 = \frac{H}{2l^2} + B\sin\phi$$

for some constant of integration B. Recall the equation for an ellipse in polar co-ordinates

$$\frac{1}{r} = \frac{1}{b} + \frac{\epsilon}{b}\sin\phi$$

Here, ϵ is the eccentricity of the ellipse: if it is zero the equation is that of a circle of radius b. In general b is the semi-latus rectum of the ellipse. If $1 > \epsilon > 0$, the closest and farthest approach to the origin are at $\frac{1}{r_{1,2}} = \frac{1}{b} \pm \frac{\epsilon}{b}$ so that the major axis is $r_2 + r_1 = \frac{2b}{1-\epsilon^2}$. So now we know the meaning of l and B in terms of the Newtonian orbital parameters.

$$b = 2r_s l^2, \quad B = \frac{\epsilon}{b}r_s$$

8.7.1 The perihelion shift

Putting

$$u = u_0 + u_1$$

to first order (we choose units with $H = 1$ for convenience)

$$u_1'' + u_1 = \frac{3}{2}u_0^2$$

$$= \frac{3}{8l^4} + \frac{3B}{2l^2}\sin\phi + \frac{3}{2}B^2\sin^2\phi$$

$$u_1'' + u_1 = \frac{3}{8l^4} + \frac{3}{4}B^2 + 3\frac{B}{2l^2}\sin\phi - \frac{3}{4}B^2\cos 2\phi$$

Although the driving terms are periodic, the solution is not periodic because of the resonant term $\sin \phi$ in the RHS.

$$u_1 = \text{periodic} + \text{constant} \phi \sin \phi$$

Proposition 8.13: In GR, the orbit is not closed. Thus GR predicts that as a planet returns to the perihelion its angle has suffered a net shift. After rewriting B, l, r_s in terms of the parameters a, ϵ, T of the orbit, the perihelion shift is found to be

$$\frac{24\pi^2 a^2}{(1 - \epsilon^2)c^2 T^2}$$

where a is the semi-major axis and T is the period of the orbit.

Exercise 8.3: Express the period of $u(\phi)$ in terms of a complete elliptic integral and hence the arithmetic geometric mean (AGM). Use this to get the perihelion shift in terms of the AGM. This perihelion shift agrees with the measured anomaly in the orbit of Mercury.

At the time Einstein proposed his theory, such a shift in the perihelion of Mercury was already known – and unexplained – for a hundred years! The prediction of GR, $43''$ of arc per century, exactly agreed with the observation: its first experimental test. For the Earth the shift of the perihelion is even smaller: $3.8''$ of arc per century. Much greater accuracy has been possible in determining the orbit of the Moon through laser ranging. The results are a quantitative vindication of GR to high precision.

Problem 8.4: The hyperboloid can also be thought of as the interior of the unit disk

$$ds^2 = \frac{dz d\bar{z}}{(1 - \bar{z}z)^2}, \quad \bar{z}z < 1$$

What are the geodesics in this description?

Problem 8.5: Consider the metric $ds^2 = \rho(y)^2[dx^2 + dy^2]$ on the upper half plane $y > 0$. Find the length of the geodesic whose end points (x, y) and (x', y) lie on a line parallel to the x-axis. Evaluate the special case $\rho(y) = y^{-1}$ corresponding to hyperbolic geometry.

Problem 8.6: Fermat's principle of optics states that the path $\mathbf{r}(s)$ of a light ray is the one that minimizes the path length $\int \frac{ds}{n(\mathbf{r}(s))}$ where s is the Euclidean arc length and $n(\mathbf{r})$ is the refractive index. Find the metric on space whose geodesics are the light rays.

Problem 8.7*: Find a metric on the group of rotations whose geodesics are the solutions of Euler's equations for a rigid body with principal moments of inertia I_1, I_2, I_3.

Problem 8.8: A Killing vector field is an infinitesimal symmetry of a metric tensor. That is, it leaves $ds^2 = g_{ij}(x)dx^i dx^j$ invariant under the infinitesimal transformation $x^i \rightarrow x^i + \epsilon v^i(x)$. Show that this leads to the differential equation

$$v^k \partial_k g_{ij} + \partial_i v^k \, g_{kj} + \partial_j v^k \, g_{ik} = 0$$

Show that corresponding to each Killing vector field is a conserved quantity, when geodesic equations are thought of as a hamiltonian system. Find these quantities for the infinitesimal rotations of the Euclidean metric.

9
Hamilton–Jacobi theory

We saw that the formulation of classical mechanics in terms of Poisson brackets allows a passage into quantum mechanics: the Poisson bracket measures the infinitesimal departure from commutativity of observables. There is also a formulation of mechanics that is connected to the Schrödinger form of quantum mechanics. Hamilton discovered this originally through the analogy with optics. In the limit of small wavelength, the wave equation (which is a second order linear PDE) becomes a first order (but non-linear) equation, called the eikonal equation. Hamilton and Jacobi found an analogous point of view in mechanics. In modern language, it is the short wavelength limit of Schrödinger's wave equation.

Apart from its conceptual value in connection with quantum mechanics, the Hamilton–Jacobi equation also provides powerful technical tools for solving problems of classical mechanics.

9.1. Conjugate variables

Recall that we got the Euler–Lagrange (E–L) equations by minimizing the action

$$S = \int_{t_1}^{t_2} L(q, \dot{q}) dt$$

over paths with fixed endpoints. It is interesting also to hold the initial point fixed and ask how the action varies as a function of the endpoint. Let us change notation slightly and call the end time t, and the variable of integration τ. Also let us call $q(t) = q$, the ending position.

$$S(t, q) = \int_{t_1}^{t} L(q(\tau), \dot{q}(\tau)) d\tau$$

From the definition of the integral, we see that

$$\frac{dS}{dt} = L$$

But,

$$\frac{dS}{dt} = \frac{\partial S}{\partial t} + \frac{\partial S}{\partial q^i} \dot{q}^i$$

so that

$$\frac{\partial S}{\partial t} = L - \frac{\partial S}{\partial q^i}\dot{q}^i$$

If we vary the path

$$\delta S = \left[\frac{\partial L}{\partial \dot{q}^i}\delta q^i\right]_{t_1}^{t} - \int_{t_1}^{t}\left[\frac{\partial L}{\partial q^i} - \frac{d}{dt}\frac{\partial L}{\partial \dot{q}^i}\right]dt$$

In deriving the E–L equations we could ignore the first term because the variation vanished at the endpoints. But looking at the dependence on the ending position, and recalling that $\frac{\partial L}{\partial \dot{q}^i} = p_i$, we get

$$\frac{\partial S}{\partial q^i} = p_i$$

Thus,

$$\frac{\partial S}{\partial t} = L - p_i\dot{q}^i$$

In other words

$$\frac{\partial S}{\partial t} = -H$$

So we see that the final values of the variables conjugate to t, q^i are given by the derivatives of S.

9.2.   The Hamilton–Jacobi equation

This allows us rewrite content of the action principle as a partial differential equation: we replace p_i by $\frac{\partial S}{\partial q^i}$ in the hamiltonian to get

$$\frac{\partial S}{\partial t} + H\left(q, \frac{\partial S}{\partial q}\right) = 0$$

Example 9.1: For the free particle $H = \frac{p^2}{2m}$ and

$$\frac{\partial S}{\partial t} + \frac{1}{2m}\left(\frac{\partial S}{\partial q}\right)^2 = 0$$

A solution to this equation is

$$S(t, q) = -Et + pq$$

for a pair of constants E, p satisfying

$$E = \frac{p^2}{2m}$$

Thus, the solution to the Hamilton–Jacobi (H–J) equation in this case is a sum of terms each depending only on one of the variables: it is *separable*. Whenever the H–J equation can be solved by such a separation of variables, we can decompose the motion into one-dimensional motions, each of which can be solved separately.

Example 9.2: The planar Kepler problem can be solved by separation of variables as well. In polar co-ordinates

$$H = \frac{p_r^2}{2m} + \frac{p_\phi^2}{2mr^2} - \frac{k}{r}$$

so that the H–J equation is

$$\frac{\partial S}{\partial t} + \frac{1}{2m}\left[\frac{\partial S}{\partial r}\right]^2 + \frac{1}{2mr^2}\left[\frac{\partial S}{\partial \phi}\right]^2 - \frac{k}{r} = 0$$

Since t, ϕ do not appear explicitly (i.e., they are *cyclic* variables), their conjugates can be assumed to be constants. So we make the ansatz

$$S(t, r, \theta, \phi) = -Et + R(r) + L\phi$$

yielding

$$\frac{1}{2m}\left[\frac{dR}{dr}\right]^2 + \frac{L^2}{2mr^2} - \frac{k}{r} = E$$

9.3. The Euler problem

Euler solved many problems in mechanics. One of them was the motion of a body under the influence of the gravitational field of two fixed bodies. This does not occur in astronomy, as the two bodies will themselves have to move under each other's gravitational field. But centuries later, exactly this problem occurred in studying the molecular ion H_2^+: an electron orbiting two protons at fixed positions. Heisenberg dusted off Euler's old method and solved its Schrödinger equation: the only exact solution of a molecule. The trick is to use a generalization of polar co-ordinates, in which the curves of constant radii are ellipses instead of circles. Place the two fixed masses at points $\pm\sigma$ along the z-axis. If r_1 and r_2 are distances of a point from these points, the potential is

$$V = \frac{\alpha_1}{r_1} + \frac{\alpha_2}{r_2}$$

with

$$r_{1,2} = \sqrt{(z \mp \sigma)^2 + x^2 + y^2}$$

We can use $\xi = \frac{r_1+r_2}{2\sigma}, \eta = \frac{r_2-r_1}{2\sigma}$ as co-ordinates.

Exercise 9.1: Note that $|\xi| \geq 1$ while $|\eta| \leq 1$. What are the surfaces where ξ is a constant and where η is a constant?

As the third co-ordinate we can use the angle ϕ of the cylindrical polar system:

$$x = \sigma\sqrt{(\xi^2 - 1)(1 - \eta^2)}\cos\phi, \quad y = \sigma\sqrt{(\xi^2 - 1)(1 - \eta^2)}\sin\phi, \quad z = \sigma\xi\eta$$

This is an orthogonal co-ordinate system, i.e., the metric is diagonal:

$$ds^2 = \sigma^2(\xi^2 - \eta^2)\left[\frac{d\xi^2}{\xi^2 - 1} + \frac{d\eta^2}{1 - \eta^2}\right] + \sigma^2(\xi^2 - 1)(1 - \eta^2)d\phi^2$$

Exercise 9.2: Prove this form of the metric. It is useful to start with the metric in cylindrical polar co-ordinates $ds^2 = d\rho^2 + \rho^2 d\phi^2 + dz^2$ and make the change of variables $\rho = \sigma\sqrt{(\xi^2 - 1)(1 - \eta^2)}$ and z as above.

Now the Lagrangian is

$$L = \frac{1}{2}m\sigma^2(\xi^2 - \eta^2)\left[\frac{\dot\xi^2}{\xi^2 - 1} + \frac{\dot\eta^2}{1 - \eta^2}\right] + \frac{1}{2}m\sigma^2(\xi^2 - 1)(1 - \eta^2)\dot\phi^2 - V(\xi, \eta)$$

leading to the hamiltonian

$$H = \frac{1}{2m\sigma^2(\xi^2 - \eta^2)}\left[(\xi^2 - 1)p_\xi^2 + (1 - \eta^2)p_\eta^2 + \left(\frac{1}{\xi^2 - 1} + \frac{1}{1 - \eta^2}\right)p_\phi^2\right] + V(\xi, \eta)$$

and the H–J equation

$$E = \frac{1}{2m\sigma^2(\xi^2 - \eta^2)}\left[(\xi^2 - 1)\left(\frac{\partial S}{\partial \xi}\right)^2 + (1 - \eta^2)\left(\frac{\partial S}{\partial \eta}\right)^2\right.$$
$$\left. + \left(\frac{1}{\xi^2 - 1} + \frac{1}{1 - \eta^2}\right)\left(\frac{\partial S}{\partial \theta}\right)^2\right] + V(\xi, \eta)$$

The potential can be written as

$$V(\xi, \eta) = -\frac{1}{\sigma}\left\{\frac{\alpha_1}{\xi - \eta} + \frac{\alpha_2}{\xi + \eta}\right\} = \frac{1}{\sigma(\xi^2 - \eta^2)}\{(\alpha_1 + \alpha_2)\xi + (\alpha_1 - \alpha_2)\eta\}$$

Since ϕ is cyclic we set $\frac{\partial S}{\partial \phi} = L$ a constant: it is the angular momentum around the axis connecting the two fixed bodies. The H–J equation becomes

$$2m\sigma^2(\xi^2 - \eta^2)E = (\xi^2 - 1)\left(\frac{\partial S}{\partial \xi}\right)^2 + 2m\sigma(\alpha_1 + \alpha_2)\xi + \frac{L^2}{\xi^2 - 1} + (1 - \eta^2)\left(\frac{\partial S}{\partial \eta}\right)^2$$
$$+ 2m\sigma(\alpha_1 - \alpha_2)\eta + \frac{L^2}{1 - \eta^2}$$

or

$$\left\{(\xi^2 - 1)\left(\frac{\partial S}{\partial \xi}\right)^2 + 2m\sigma(\alpha_1 + \alpha_2)\xi + \frac{L^2}{\xi^2 - 1} + 2mE\sigma^2(\xi^2 - 1)\right\}$$

$$+ \left\{(1 - \eta^2)\left(\frac{\partial S}{\partial \eta}\right)^2 + 2m\sigma(\alpha_1 - \alpha_2)\eta + \frac{L^2}{1 - \eta^2} + 2mE\sigma^2(1 - \eta^2)\right\} = 0$$

This suggests the separation of variables

$$S = A(\xi) + B(\eta)$$

where each satisfies the ODE

$$(\xi^2 - 1)A'^2 + 2m\sigma(\alpha_1 + \alpha_2)\xi + \frac{L^2}{\xi^2 - 1} + 2mE\sigma^2(\xi^2 - 1) = K$$

$$(1 - \eta^2)B'^2 + 2m\sigma(\alpha_1 - \alpha_2)\eta + \frac{L^2}{1 - \eta^2} + 2mE\sigma^2(1 - \eta^2) = -K$$

The solutions are elliptic integrals.

9.4. The classical limit of the Schrödinger equation*

Recall that the Schrödinger equation of a particle in a potential is

$$-\frac{\hbar^2}{2m}\nabla^2\psi + V\psi = i\hbar\frac{\partial \psi}{\partial t}$$

In the limit of small $\hbar$ (i.e., when quantum effects are small) this reduces to the H–J equation. The idea is to make the change of variables

$$\psi = e^{\frac{i}{\hbar}S}$$

so that the equation becomes

$$-\frac{i\hbar}{2m}\nabla^2 S + \frac{1}{2m}(\nabla S)^2 + V + \frac{\partial S}{\partial t} = 0$$

If we ignore the first term we get the H–J equation.

Co-ordinate systems and potentials in which the H–J is separable also allow the solution of the Schrödinger equation by separation of variables. A complete list is given in Landau and Lifshitz (1977).

9.5.   Hamilton–Jacobi equation in Riemannian manifolds*

Given any metric $ds^2 = g_{ij}dq^i dq^j$ in configuration space, we have the Lagrangian

$$L = \frac{1}{2}mg_{ij}dq^i dq^j - V(q)$$

The momenta are

$$p_i = mg_{ij}\dot{q}^j$$

and the hamiltonian is

$$H = \frac{1}{2}g^{ij}p_i p_j + V(q)$$

The Hamilton–Jacobi equation becomes, for a given energy

$$\frac{1}{2m}g^{ij}\frac{\partial S}{\partial q^i}\frac{\partial S}{\partial q^j} + V = E$$

If the metric is diagonal, ("orthogonal co-ordinate system") the inverse is easier to calculate.

In the absence of a potential this becomes

$$g^{ij}\frac{\partial S}{\partial q^i}\frac{\partial S}{\partial q^j} = \text{constant}$$

which is the H–J version of the geodesic equation.

Even when there is a potential, we can rewrite this as

$$\tilde{g}^{ij}\frac{\partial S}{\partial q^i}\frac{\partial S}{\partial q^j} = 1, \quad \tilde{g}^{ij} = \frac{1}{2m[E - V(q)]}$$

Thus the motion of a particle in a potential can be thought of as geodesic motion in an effective metric

$$d\tilde{s}^2 = 2m[E - V(q)]g_{ij}dq^i dq^j$$

This is related to the Maupertuis principle we discussed earlier. Note that only the classically allowed region $E > V(q)$ is accessible to these geodesics.

9.6. Analogy to optics*

Light is a wave that propagates at a constant speed c in the vacuum. In a medium, its effective speed depends on the refractive index. The wave equation of light of wavenumber k in a medium whose refractive index varies with position is

$$k^2 n^2(\mathbf{r})\psi + \nabla^2\psi = 0.$$

When $n = 1$ a typical solution of this equation is a plane wave $e^{i\mathbf{k}\cdot\mathbf{r}}$. The wavelength of light is $\lambda = \frac{2\pi}{k}$. For visible light $\lambda \sim 10^{-6}$ m. In most cases, the refractive index is almost a constant over such a small distance. Thus we can often make an approximation that simplifies the wave equation. To set the stage, change variables

$$\psi = e^{ikS}$$

to get

$$(\nabla S)^2 = n^2(\mathbf{r}) - \frac{i}{k}\nabla^2 S$$

For small wavelength (large k) we can ignore the last term, so that the wave equation reduces to the *eikonal equation*:

$$(\nabla S)^2 = n^2(\mathbf{r}).$$

Mathematically, this is identical to the Hamilton–Jacobi equation of a mechanical system, with the correspondence

$$2m[E - V(\mathbf{r})] = n^2(\mathbf{r})$$

Thus, methods of classical mechanics (variational principles, conservation laws, Poisson brackets) can be used to aid the ancient art of ray tracing, which is still used to design optical instruments. The path of a light ray can be thought of as the geodesics of the Fermal metric

$$ds^2_{\text{Fermat}} = \frac{1}{n^2(\mathbf{r})} d\mathbf{r} \cdot d\mathbf{r}$$

The analogy of the optical wave equation to the Schrödinger equation has its uses as well: quantum mechanical methods can be used to design waveguides, as Julian Schwinger did during the war. There is even an analogue of the uncertainty principle in optics. Using the time it takes for a radio signal to bounce back from an object we can determine its position (radar). From the shift in the frequency of the returning signal we can deduce its speed (Doppler radar). The uncertainty in position times the uncertainty in speed cannot be less than the wavelength times the speed of light.

Problem 9.3: An isotropic top is a system whose configuration space is the unit sphere. It has hamiltonian

$$H = \frac{1}{2I}\mathbf{L}^2 + mgl\cos\theta$$

where I is the moment of inertia. Also, $\mathbf{L}$ is angular momentum, which generates rotations on the sphere through canonical transformations. Show that, in polar co-ordinates,

$$\mathbf{L}^2 = p_\phi^2 + \frac{p_\theta^2}{\sin^2\theta}$$

Obtain the Hamilton–Jacobi equations of this system in these co-ordinates. Show that it is separable. Use that to determine the solutions of the equations of the top.

Problem 9.4: Show that the H–J equation can be solved by separation of variables

$$S(t, r, \theta, \phi) = T(t) + R(r) + \Theta(\theta) + \Phi(\phi)$$

in spherical polar co-ordinates for any potential of the form $V(r, \theta, \phi) = a(r) + \frac{b(\theta)}{r^2} + \frac{c(\phi)}{r^2\sin^2\theta}$. The Kepler problem is a special case of this.

Problem 9.5: Show that the Schrödinger equation of the H_2^+ molecular ion can be separated in elliptical co-ordinates.

10
Integrable systems

In quantum mechanics with a finite number of independent states, the hamiltonian is a Hermitian matrix. It can be diagonalized by a unitary transformation. The analog in classical physics is a canonical transformation that brings the hamiltonian to *normal form*: so that it depends only on variables P_i that commute with each other. In this form Hamilton's equations are trivial to solve:

$$\frac{dQ^i}{dt} = \frac{\partial H}{\partial P_i}, \quad \frac{dP_i}{dt} = 0, \quad i = 1, \cdots, n \tag{10.1}$$

$$P_i(t) = P_i(0), \quad Q^i(t) = Q^i(0) + \omega^i(P)t, \quad \omega^i(P) = \frac{\partial H}{\partial P_i}$$

Thus, the whole problem of solving the equations of motion amounts to finding a canonical transformation $p_i(P, Q), q^i(P, Q)$ such that $H(p(P, Q), q(P, Q))$ only depends on P_i. If we knew the generating function $S(P, q)$ we could determine the canonical transformation by solving

$$p_i = \frac{\partial S(P, q)}{\partial q^i}, \quad Q^i = \frac{\partial S(P, q)}{\partial P_i} \tag{10.2}$$

So how do we determine $S(P, q)$? If we put the first of the above equations 10.2 into the hamiltonian

$$H\left(\frac{\partial S}{\partial q}, q\right) = E \tag{10.3}$$

This is the Hamilton–Jacobi equation. If we can solve this (for example, by the method of separation of variables explained below) we can then solve 10.2 and we would have brought the hamiltonian to normal form. The conserved quantities P appear as constants of integration (separation constants) in solving the H–J equation.

In most cases of interest, the orbits of the system lie within some bounded region of the phase space. Then the variables Q^i are angles. The orbits lie on a torus whose co-ordinates are Q^i, determined by the conserved quantities P. Conserved quantities are also called invariants, so these are the invariant tori.

Example 10.1: The simplest integrable system is of one degree of freedom and

$$H(P) = \omega P$$

for a constant frequency ω. The variable Q is an angle with range $[0, 2\pi]$. The solution to the H–J equation $\omega \frac{\partial S}{\partial Q} = E$ is

$$S(Q) = \frac{E}{\omega} Q = PQ$$

It generates the identity transformation because the system is already in normal form. Notice that as the angle changes from 0 to 2π the action changes by $2\pi P$. This is a convenient normalization of P which we will try to achieve in other examples.

10.1.   The simple harmonic oscillator

This is the prototype of an integrable system. We choose units such that the mass is equal to unity.

$$H = \frac{p^2}{2} + \frac{1}{2}\omega^2 q^2$$

The orbits are ellipses in phase space.

$$q(t) = \frac{\sqrt{2E}}{\omega} \cos \omega t, \quad p(t) = \sqrt{2E} \sin \omega t$$

This suggests that we choose as the position variable

$$Q = \arctan \frac{\omega q}{p}$$

since it evolves linearly in time.

Its conjugate variable is

$$P = \frac{1}{\omega}\left[\frac{p^2}{2} + \frac{1}{2}\omega^2 q^2\right]$$

Exercise 10.1: Verify that $\{P, Q\} = 1$. Recall that the Poisson bracket is the Jacobian determinant in two dimensions, so what you need to show is that $dpdq = dPdQ$.

Solution
One way to see this quickly is to recall that if we go from Cartesian to polar co-ordinates

$$dxdy = d\left[\frac{r^2}{2}\right]d\theta$$

Now put $x = \omega q, y = p, Q = \theta$.
Thus we can write the hamiltonian in normal form

$$H = \omega P$$

10.1.1   Solution using the H–J equation

The Hamilton–Jacobi equation is

$$\frac{1}{2}\left[S'^2(q) + \omega^2 q^2\right] = E$$

The solution is

$$S(q) = \frac{1}{2}q\sqrt{2E - q^2\omega^2} + \frac{E}{\omega}\arctan\left(\frac{q\omega}{\sqrt{2E - q^2\omega^2}}\right)$$

In one orbital period, this $S(q)$ increases by $2\pi\frac{E}{\omega}$. This suggests that we choose

$$P = \frac{E}{\omega}$$

and

$$S(P, q) = \frac{1}{2}q\sqrt{2\omega P - q^2\omega^2} + P\arctan\left(\frac{q\omega}{\sqrt{2\omega P - q^2\omega^2}}\right)$$

Differentiating, 10.2 becomes

$$p = \sqrt{\omega(2P - \omega q^2)}, \quad Q = \arctan\frac{\omega q}{p}$$

which is exactly what we had above.

10.2.   The general one-dimensional system

Consider now any hamiltonian system with one degree of freedom

$$H = \frac{1}{2m}p^2 + V(q)$$

Assume for simplicity that the curves of constant energy are closed, i.e., that the motion is periodic. Then we look for a co-ordinate system Q, P in the phase space such that the period of Q is 2π and which is canonical

$$dpdq = dPdQ$$

In this case the area enclosed by the orbit of a fixed value of P will be just $2\pi P$. On the other hand, this area is just $\oint_H pdq$ over a curve constant energy in the original co-ordinates (Stokes' theorem). Thus we see that

$$P = \frac{1}{2\pi}\oint_H pdq = \frac{1}{\pi}\int_{q_1(H)}^{q_2(H)}\sqrt{2m\left[H - V(q)\right]}dq$$

where $q_{1,2}(H)$ are *turning points*, i.e., the roots of the equation $H - V(q) = 0$.

If we can evaluate this integral, we will get P as a function of H. Inverting this function will give H as a function of P, which is its normal form.

By comparing with the H–J equation, we see that P is simply $\frac{1}{2\pi}$ times the change in the action over one period of the q variable (i.e., from q_1 to q_2 and back again to q_1):

$$\frac{1}{2m}\left[\frac{\partial S}{\partial q}\right]^2 + V(q) = H, \quad S(P,q) = \int_{q_1}^{q} \sqrt{2m[H(P) - V(q)]}dq$$

This is why P is called the *action variable*. Its conjugate, *the angle variable Q*, is now given by

$$Q = \frac{\partial S(P,q)}{\partial P} = \sqrt{\frac{m}{2}}\frac{\partial H(P)}{\partial P}\int_{q_1}^{q}\frac{dq}{\sqrt{[H(P) - V(q)]}} \tag{10.4}$$

10.3.   Bohr–Sommerfeld quantization

Once the hamiltonian is brought to normal form, there is a natural way to quantize the system that explicitly displays its energy eigenvalues. The Schrödinger equation becomes

$$H\left(-i\hbar\frac{\partial}{\partial Q}\right)\psi = E\psi$$

The solution is a "plane wave"

$$\psi = e^{\frac{i}{\hbar}PQ}$$

If the Q-variables are angles (as in the SHO) with period 2π we must require that $P = \hbar n$ for $n = 0, 1, 2 \cdots$ in order that the wave function is single-valued. Thus the spectrum of the quantum hamiltonian is

$$E_n = H(\hbar n)$$

In the case of the SHO, we get this way

$$E_n = \hbar\omega n, \quad n = 0, 1, \cdots$$

This is almost the exact answer we would get by solving the Schrödinger equation in terms of q: only an additive constant $\frac{1}{2}\hbar\omega$ is missing.

Using the above formula for P, we see that the quantization rule for energy can be expressed as

$$\oint pdq = 2\pi\hbar n$$

This is known as Bohr–Sommerfeld (B–S) quantization and provides a semi-classical approximation to the quantum spectrum. In some fortuitous cases (such as the SHO or the hydrogen atom) it gives almost the exact answer.

10.4. The Kepler problem

The action-angle variables of the Kepler problem were found by Delaunay. We already know that p_ϕ and $L^2 = p_\theta^2 + \frac{p_\phi^2}{\sin^2\theta}$ (the component of angular momentum in some direction, say z, and the total angular momentum) are a pair of commuting conserved quantities. So this reduces to a problem with just one degree of freedom.

$$H = \frac{p_r^2}{2m} + \frac{L^2}{2mr^2} - \frac{k}{r}$$

To find the normal form we need to evaluate the integral

$$P = \frac{1}{\pi} \int_{r_1}^{r_2} \sqrt{2m\left[H - \frac{L^2}{2mr^2} + \frac{k}{r}\right]}\, dr$$

between turning points. This is equal to (see below)

$$P = -L - \frac{\sqrt{2mk}}{2\sqrt{-H}}$$

Thus

$$H(P, L) = -\frac{mk^2}{2(P + L)^2}$$

Within the B–S approximation, the action variables P, L are both integers.

So the quantity $P + L$ has to be an integer in the B–S quantization: it is the *principal quantum number* of the hydrogenic atom and the above formula gives its famous spectrum. If we include the effects of special relativity, the spectrum depends on P, L separately, not just on the sum: this is the *fine structure* of the hydrogenic atom.

10.4.1   A contour integral*

It is possible to evaluate the integral by trigonometric substitutions, but it is a mess. Since we only want the integral between turning points, there is a trick involving contour integrals. Consider the integral $\oint f(z)dz$ over a counter-clockwise contour of the function

$$f(z) = \frac{\sqrt{Az^2 + Bz + C}}{z}$$

On the Riemann sphere, $f(z)dz$ has a branch cut along the line connecting the zeros of the quadratic under the square roots. It has a simple pole at the origin. This integrand also has a pole at infinity; this is clear if we transform to $w = \frac{1}{z}$

$$\frac{\sqrt{Az^2 + Bz + C}}{z}dz = -\frac{\sqrt{Cw^2 + Bw + A}}{w^2}dw$$

The residue of the pole $w = 0$ is

$$-\frac{B}{2\sqrt{A}}$$

The integral $\oint f(z)dz$ over a contour that surrounds all of these singularities must be zero: it can be shrunk to some point on the Riemann sphere. So the sum of the residues on the two simple poles plus the integral of the discontinuity across the branchcut must be zero:

$$2\int_{z_1}^{z_2} \frac{\sqrt{Az^2 + Bz + C}}{z} dz = 2\pi i \left[\sqrt{C} - \frac{B}{2\sqrt{A}}\right]$$

With the choice

$$A = 2mH, \quad B = 2mk, \quad C = -L^2$$

we get

$$\int_{r_1}^{r_2} \sqrt{2m\left[H - \frac{L^2}{2mr^2} + \frac{k}{r}\right]} \, dr = i\pi \left[\sqrt{-L^2} - \frac{2mk}{2\sqrt{(2mH)}}\right]$$

$$= \pi \left[-L - \frac{\sqrt{2m}k}{2\sqrt{-H}}\right]$$

10.4.2   Delaunay variable*

What is the angle variable Q conjugate to the action variable P? This is determined by applying the formula 10.4 found above:

$$Q = \sqrt{\frac{m}{2}} \frac{\partial H(P, L)}{\partial P} \int_{r_1}^{r} \frac{dr}{\sqrt{\left[H(P, L) - \frac{L^2}{2mr^2} + \frac{k}{r}\right]}}$$

To evaluate the integral it is useful to make the change of variable

$$H(P, L) - \frac{L^2}{2mr^2} + \frac{k}{r} = \frac{mkL}{2(P + L)^2}\left(\frac{\epsilon \sin \chi}{1 - \epsilon \cos \chi}\right)^2$$

This variable χ is called the eccentric anomaly, for historical reasons. It vanishes at the turning point r_1 and is equal to π at the other turning point r_2. The eccentricity of the orbit is

$$\epsilon = \left[1 + \frac{2L^2 H(P, L)}{mk^2}\right]^{\frac{1}{2}} = \sqrt{1 - \frac{L^2}{(P + L)^2}}$$

After some calculation, the integral evaluates to the Delaunay angle variable

$$Q = \chi - \epsilon \sin \chi.$$

10.5.   The relativistic Kepler problem*

Sommerfeld worked out the effect of special relativity on the Kepler problem, which explained the fine structure of the hydrogen atom within the Bohr model. A similar calculation can also be done for the general relativistic problem, but as yet it does not have a physical realization: gravitational effects are negligible in the atom, as are quantum effects in planetary dynamics. We start with the relation of momentum to energy in special relativity for a free particle:

$$p_t^2 - c^2\mathbf{p}^2 = m^2c^4$$

In the presence of an electrostatic potential this is modified to

$$[p_t - eV(r)]^2 - c^2\mathbf{p}^2 = m^2c^4$$

In spherical polar co-ordinates

$$[p_t - eV(r)]^2 - c^2p_r^2 - \frac{c^2L^2}{r^2} = m^2c, \quad L^2 = p_\theta^2 + \frac{p_\phi^2}{\sin^2\theta}$$

Since p_t, L, p_ϕ are still commuting quantities, this still reduces to a one-dimensional problem. So we still define

$$P = \frac{1}{2\pi}\oint p_r\,dr$$

as before. With the Coulomb potential $eV(r) = -\frac{k}{r}$ we again have a quadratic equation for p_r. The integral can be evaluated by the contour method again.

> **Exercise 10.2:** Derive the relativistic formula for the spectrum of the hydrogen atom by applying the Bohr–Sommerfeld quantization rule.

10.6.   Several degrees of freedom

As long as the H–J equation is separable, there is a generalization of the above procedure to a system with several degrees of freedom.

$$S = \sum_i S_i(q_i)$$

$$H = \sum_i H_i\left(q_i, \frac{\partial S_i}{\partial q_i}\right)$$

$$H_i\left(q_i, \frac{\partial S_i}{\partial q_i}\right) + \frac{\partial S_i}{\partial t} = 0$$

In essence, separation of variables breaks up the system into decoupled one-dimensional systems, each of which can be solved as above. This is essentially what we did when we dealt with the Kepler problem above. The momentum ("action") variables are the integrals

$$P_i = \frac{1}{2\pi} \oint p_i \, dq_i$$

10.7. The heavy top

We solved the motion of the rigid body on which no torque is acting. Some cases with torque are integrable as well. An interesting example is a top: a rigid body on which gravity exerts a torque. Recall that angular momentum and angular velocity are related by $\mathbf{L} = I\boldsymbol{\Omega}$, where I is the moment of inertia. It is again convenient to go to the (non-inertial) body-centered frame in which I is diagonal, so that $L_1 = I_1\Omega_1$ etc. The equation of motion is then

$$\left[\frac{d\mathbf{L}}{dt}\right]_{\text{inertial}} \equiv \frac{d\mathbf{L}}{dt} + \boldsymbol{\Omega} \times \mathbf{L} = \mathbf{T}$$

The torque is

$$\mathbf{T} = mg\mathbf{R} \times \mathbf{n}$$

where $\mathbf{R}$ is the vector connecting the point supporting the weight of the top to its center of mass. Also, g the magnitude of the acceleration due to gravity and $\mathbf{n}$ the unit vector along its direction. In the body-centered frame, $\mathbf{R}$ is a constant, while $\mathbf{n}$ varies with time. Since gravity has a fixed direction in the inertial frame

$$\left[\frac{d\mathbf{n}}{dt}\right]_{\text{inertial}} \equiv \frac{d\mathbf{n}}{dt} + \boldsymbol{\Omega} \times \mathbf{n} = 0$$

So the equations to be solved are (with $a_1 = \frac{1}{I_2} - \frac{1}{I_3}$ etc. as before)

$$\frac{dL_1}{dt} + a_1 L_2 L_3 = mg(R_2 n_3 - R_3 n_2)$$

$$\frac{dn_1}{dt} + \frac{1}{I_2} L_2 n_3 - \frac{1}{I_3} L_3 n_2 = 0$$

and cyclic permutations thereof. These equations follow from the hamiltonian

$$H = \frac{L_1^2}{2I_1} + \frac{L_2^2}{2I_2} + \frac{L_3^2}{2I_3} + mg\mathbf{R} \cdot \mathbf{n}$$

and Poisson brackets

$$\{L_1, L_2\} = L_3, \quad \{L_1, n_2\} = n_3, \quad \{n_1, n_2\} = 0$$

and cyclic permutations. These simply say that $\mathbf{n}$ transforms as a vector under canonical transformations (rotations) generated by $\mathbf{L}$ and that the components of $\mathbf{n}$ commute with each other. Obviously, H and $\mathbf{n}^2$ are conserved.

10.7.1 Lagrange top

The general case is not exactly solvable: the motion is chaotic. Lagrange discovered that the case where the body has a symmetry around some axis (as a toy top does) is solvable. If the symmetry is around the third axis $I_1 = I_2$, so that $a_3 = 0, a_1 = -a_2$. Also, assume that the point of support of the weight lies along the third axis. Then the center of mass will be at $\mathbf{R} = (0, 0, R)$. It follows that L_3 is conserved. The hamiltonian simplifies to

$$H = \frac{L^2}{2I_1} + mgRn_3 + L_3^2 \left(\frac{1}{2I_3} - \frac{1}{2I_1} \right)$$

The last term is constant and commutes with the other two terms. If we introduce polar co-ordinates on the unit sphere

$$n_3 = \cos \theta, \quad n_1 = \sin \theta \cos \phi, \quad n_2 = \sin \theta \sin \phi$$

a canonical representation for the commutation relations would be given by

$$L_3 = p_\phi, \quad L_1 = \sin \phi p_\theta + \cot \theta \cos \phi p_\phi, \quad L_2 = -\cos \phi p_\theta + \cot \theta \sin \phi p_\phi$$

and the hamiltonian becomes

$$H = \frac{p_\theta^2}{2I_1} + \frac{p_\phi^2}{2I_1 \sin^2 \theta} + mgR \cos \theta + \left(\frac{1}{2I_3} - \frac{1}{2I_1} \right) p_\phi^2$$

Since p_ϕ is a constant, this reduces to a particle moving on a circle with the effective potential

$$V(\theta) = \frac{p_\phi^2}{2I_1 \sin^2 \theta} + mgR \cos \theta$$

It oscillates around the stable equilibrium point of this potential.

Problem 10.3: Express the orbits of the Lagrange top explicitly in terms of elliptic functions. Transform back to the inertial co-ordinate system. Plot some examples, to show the *nutations* of the axis of rotation of the top.

Problem 10.4: Suppose that the top is symmetric around an axis (so that $I_1 = I_2$) but the point of support is not along that axis. Instead, suppose that the vector to the center of mass $\mathbf{R} = (R, 0, 0)$ is in a direction normal to the symmetry axis. This is not integrable in general. Kovalevskaya showed that if in addition $I_1 = I_2 = 2I_3$ the system is integrable. Her key discovery was that

$$K = |\xi|^2, \quad \xi = \frac{(L_1 + iL_2)^2}{2I_1} - mgR(n_1 + in_2)$$

is conserved. This is a surprise, as it does not follow from any obvious symmetry. Prove this fact by direct calculation of its time derivative. Use K to complete the solution of the Kovalevskaya top.

Problem 10.5: Find the spectrum of the hydrogen molecular ion H_2^+ within the Bohr–Sommerfeld approximation. (Use elliptic polar co-ordinates to separate the H–J equation as in the last chapter. Express the action variables in terms of complete elliptic integrals.)

11
The three body problem

Having solved the two body problem, Newton embarked on a solution of the three body problem: the effect of the Sun on the orbit of the Moon. It defeated him. The work was continued by many generations of mathematical astronomers: Euler, Lagrange, Airy, Hamilton, Jacobi, Hill, Poincaré, Kolmogorov, Arnold, Moser ... It still continues. The upshot is that it is not possible to solve the system in "closed form": more precisely that the solution is not a known analytic function of time. But a solution valid for fairly long times was found by perturbation theory around the two body solution: the series will eventually break down as it is only asymptotic and not convergent everywhere. There are regions of phase space where it converges, but these regions interlace those where it is divergent. The problem is that there are resonances whenever the frequencies of the unperturbed solution are rational multiples of each other.

The most remarkable result in this subject is a special exact solution of Lagrange: there is a stable solution in which the three bodies revolve around their center of mass, keeping their positions at the vertices of an equilateral triangle. This solution exists even when the masses are not equal; i.e., the three-fold symmetry of the equilateral triangle holds even if the masses are not equal! Lagrange thought that such special orbits would not appear in nature. But we now know that Jupiter has captured some asteroids (Trojans) into such a resonant orbit. Recently, it was found that the Earth also has such a co-traveller at one of its Lagrange points.

The theory is mainly of mathematical (conceptual) interest these days as it is easy to solve astronomical cases numerically. As the first example of a chaotic system, the three body problem remains fascinating to mathematicians. New facts are still being discovered. For example, Simo, Chenciner, Montgomery found a solution ("choreography") in which three bodies of equal mass follow each other along a common orbit that has the shape of a figure eight.

11.1.   Preliminaries

Let $\mathbf{r}_a$, for $a = 1, 2, \cdots, n$ be the positions of n bodies interacting through the gravitational force. The Lagrangian is

$$ L = \frac{1}{2} \sum_a m_a \dot{\mathbf{r}}_a^2 - U, \quad U = -\sum_{a<b} \frac{G m_a m_b}{|\mathbf{r}_a - \mathbf{r}_b|} $$

Immediately we note the conservation laws of energy (hamiltonian)

$$H = \sum_a \frac{\mathbf{p}_a^2}{2m_a} + U, \quad \mathbf{p}_a = m\dot{\mathbf{r}}_a$$

total momentum

$$\mathbf{P} = \sum_a \mathbf{p}_a$$

and angular momentum

$$\mathbf{L} = \sum_a \mathbf{r}_a \times \mathbf{p}_a$$

11.2.   Scale invariance

Another symmetry is a scale invariance. If $\mathbf{r}_a(t)$ is a solution, so is

$$\lambda^{-\frac{2}{3}} \mathbf{r}_a(\lambda t)$$

Under this transformation,

$$\mathbf{p}_a \to \lambda^{\frac{1}{3}} \mathbf{p}_a, \quad H \to \lambda^{\frac{2}{3}} H, \quad \mathbf{L} \to \lambda^{-\frac{1}{3}} \mathbf{L}$$

In the two body problem, this leads to Kepler's scaling law $T^2 \propto R^3$ relating period to the semi-major axis of the ellipse. There is no conserved quantity corresponding to this symmetry, as it does not leave the Poisson brackets invariant. But it does lead to an interesting relation for the moment of inertia about the center of mass

$$I = \frac{1}{2} \sum_a m_a \mathbf{r}_a^2$$

Clearly,

$$\frac{dI}{dt} = \sum_a \mathbf{r}_a \cdot \mathbf{p}_a \equiv D$$

It is easily checked that D

$$\{D, \mathbf{r}_a\} = \mathbf{r}_a, \quad \{D, \mathbf{p}_a\} = -\mathbf{p}_a$$

So that

$$\{D, T\} = -2T$$

for kinetic energy and

$$\{D, U\} = -U$$

for potential energy. In other words

$$\{D, H\} = -2T - U = -2H + U$$

That is

$$\frac{dD}{dt} = 2H - U$$

or

$$\frac{d^2 I}{dt^2} = 2H - U$$

If the potential had been proportional to the inverse square distance (unlike the Newtonian case) this would have said instead

$$\ddot{I} = 2H.$$

We will return to this case later.

11.3. Jacobi co-ordinates

Recall that the Lagrangian of the two body problem

$$L = \frac{1}{2} m_1 \dot{\mathbf{r}}_1^2 + \frac{1}{2} m_2 \dot{\mathbf{r}}_2^2 - U(|\mathbf{r}_1 - \mathbf{r}_2|)$$

can be written as

$$L = \frac{1}{2} M_1 \dot{\mathbf{R}}_1^2 - U(|\mathbf{R}_1|) + \frac{1}{2} M_2 \dot{\mathbf{R}}_2^2$$

where

$$\mathbf{R}_1 = \mathbf{r}_2 - \mathbf{r}_1, \quad \mathbf{R}_2 = \frac{m_1 \mathbf{r}_1 + m_2 \mathbf{r}_2}{m_1 + m_2}$$

and

$$M_1 = \frac{m_1 m_2}{m_1 + m_2}, \quad M_2 = m_1 + m_2$$

This separates the center of mass (c.m.) co-ordinate $\mathbf{R}_2$ from the relative co-ordinate $\mathbf{R}_1$.

Jacobi found a generalization to three particles:

$$\mathbf{R}_1 = \mathbf{r}_2 - \mathbf{r}_1, \quad \mathbf{R}_2 = \mathbf{r}_3 - \frac{m_1\mathbf{r}_1 + m_2\mathbf{r}_2}{m_1 + m_2}, \quad \mathbf{R}_3 = \frac{m_1\mathbf{r}_1 + m_2\mathbf{r}_2 + m_3\mathbf{r}_3}{m_1 + m_2 + m_3}$$

$\mathbf{R}_2$ is the position of the third particle relative to the c.m. of the first pair.
   The advantage of this choice is that the kinetic energy

$$T = \frac{1}{2}m_1\dot{\mathbf{r}}_1^2 + \frac{1}{2}m_2\dot{\mathbf{r}}_2^2 + \frac{1}{2}m_3\dot{\mathbf{r}}_3^2$$

remains diagonal (i.e., no terms such as $\dot{\mathbf{R}}_1 \cdot \dot{\mathbf{R}}_2$):

$$T = \frac{1}{2}M_1\dot{\mathbf{R}}_1^2 + \frac{1}{2}M_2\dot{\mathbf{R}}_2^2 + \frac{1}{2}M_3\dot{\mathbf{R}}_3^2$$

with

$$M_1 = \frac{m_1 m_2}{m_1 + m_2}, \quad M_2 = \frac{(m_1 + m_2)m_3}{m_1 + m_2 + m_3}, \quad M_3 = m_1 + m_2 + m_3$$

Moreover

$$\mathbf{r}_2 - \mathbf{r}_3 = \mu_1\mathbf{R}_1 - \mathbf{R}_2, \quad \mathbf{r}_1 - \mathbf{r}_3 = -\mu_2\mathbf{R}_1 - \mathbf{R}_2$$

with

$$\mu_1 = \frac{m_1}{m_1 + m_2}, \quad \mu_1 + \mu_2 = 1$$

This procedure has a generalization to an arbitrary number of bodies.

Exercise 11.1: The construction of Jacobi co-ordinates is an application of Gram–Schmidt orthogonalization, a standard algorithm of linear algebra. Let the mass matrix be $m = \begin{pmatrix} m_1 & 0 & 0 \\ 0 & m_2 & 0 \\ 0 & 0 & m_3 \end{pmatrix}$. Starting with $\mathbf{R}_1 = \mathbf{r}_2 - \mathbf{r}_1$, find a linear combination $\mathbf{R}_2$ such that $\mathbf{R}_2^T \cdot m\mathbf{R}_1 = 0$. Then find $\mathbf{R}_3$ such that $\mathbf{R}_3^T m\mathbf{R}_1 = 0 = \mathbf{R}_3^T m\mathbf{R}_1$. Apply the linear transformation $\mathbf{R}_a = L_{ab}\mathbf{r}_a$ to get the reduced masses $M = L^T m L$. Because of orthogonality, it will be diagonal $M = \begin{pmatrix} M_1 & 0 & 0 \\ 0 & M_2 & 0 \\ 0 & 0 & M_3 \end{pmatrix}$.
   Thus, the Lagrangian of the three body problem with pairwise central potentials

$$L = \frac{1}{2}m_1\dot{\mathbf{r}}_1^2 + \frac{1}{2}m_2\dot{\mathbf{r}}_2^2 + \frac{1}{2}m_3\dot{\mathbf{r}}_3^2 - U_{12}(|\mathbf{r}_1 - \mathbf{r}_2|) - U_{13}(|\mathbf{r}_1 - \mathbf{r}_3|) - U_{23}(|\mathbf{r}_2 - \mathbf{r}_3|)$$

becomes

$$L = \frac{1}{2}M_1\dot{\mathbf{R}}_1^2 + \frac{1}{2}M_2\dot{\mathbf{R}}_2^2 - U_{12}(|\mathbf{R}_1|) - U_{13}(|\mathbf{R}_2 + \mu_2\mathbf{R}_1|) - U_{23}(|\mathbf{R}_2 - \mu_1\mathbf{R}_1|)$$
$$+ \frac{1}{2}M_3\dot{\mathbf{R}}_3^2$$

Again the c.m. co-ordinate $\mathbf{R}_3$ satisfies

$$\ddot{\mathbf{R}}_3 = 0$$

So we can pass to a reference frame in which it is at rest, and choose the origin at the c.m.:

$$\mathbf{R}_3 = 0$$

Thus the Lagrangian reduces to

$$L = \frac{1}{2}M_1\dot{\mathbf{R}}_1^2 + \frac{1}{2}M_2\dot{\mathbf{R}}_2^2 - U_{12}(|\mathbf{R}_1|) - U_{13}(|\mathbf{R}_2 + \mu_2\mathbf{R}_1|) - U_{23}(|\mathbf{R}_2 - \mu_1\mathbf{R}_1|)$$

The hamiltonian is

$$H = \frac{\mathbf{P}_1^2}{2M_1} + \frac{\mathbf{P}_2^2}{2M_2} + U_{12}(|\mathbf{R}_1|) + U_{13}(|\mathbf{R}_2 + \mu_2\mathbf{R}_1|) + U_{23}(|\mathbf{R}_2 - \mu_1\mathbf{R}_1|)$$

The total angular momentum

$$\mathbf{L} = \mathbf{R}_1 \times \mathbf{P}_1 + \mathbf{R}_2 \times \mathbf{P}_2$$

is conserved as well.

11.3.1   Orbits as geodesics

The Hamilton–Jacobi equation becomes

$$\frac{1}{2M_1}\left[\frac{\partial S}{\partial \mathbf{R}_1}\right]^2 + \frac{1}{2M_2}\left[\frac{\partial S}{\partial \mathbf{R}_2}\right]^2 + U_{12}(|\mathbf{R}_1|) + U_{13}(|\mathbf{R}_2 + \mu_2\mathbf{R}_1|) + U_{23}(|\mathbf{R}_2 - \mu_1\mathbf{R}_1|) = E$$

Or,

$$[E - \{U_{12}(|\mathbf{R}_1|) + U_{13}(|\mathbf{R}_2 + \mu_2\mathbf{R}_1|)$$
$$+ U_{23}(|\mathbf{R}_2 - \mu_1\mathbf{R}_1|)\}]^{-1}\left\{\frac{1}{2M_1}\left[\frac{\partial S}{\partial \mathbf{R}_1}\right]^2 + \frac{1}{2M_2}\left[\frac{\partial S}{\partial \mathbf{R}_2}\right]^2\right\} = 1$$

This describes geodesics of the metric

$$ds^2 = [E - \{U_{12}(|\mathbf{R}_1|) + U_{13}(|\mathbf{R}_2 + \mu_2\mathbf{R}_1|) + U_{23}(|\mathbf{R}_2 - \mu_1\mathbf{R}_1|)\}]\left\{M_1 d\mathbf{R}_1^2 + M_2 d\mathbf{R}_2^2\right\}$$

The curvature of this metric ought to give insights into the stability of the three body problem. Much work can still be done in this direction.

In the special case $E = 0, U(r) \propto \frac{1}{r}$, this metric has a scaling symmetry: $\mathbf{R}_a \to \lambda \mathbf{R}_a$, $ds^2 \to \lambda ds^2$. If $E \neq 0$ we can use this symmetry to set $E = \pm 1$, thereby choosing a unit of time and space as well.

11.4.   The $\frac{1}{r^2}$ potential

If the potential were $\frac{1}{r^2}$ and not $\frac{1}{r}$ as in Newtonian gravity, dilations would be a symmetry. Since we are interested in studying the three body problem as a model of chaos and not just for astronomical applications, we could pursue this simpler example instead. Poincaré initiated this study in 1897, as part of his pioneering study of chaos. Montgomery has obtained interesting new results in this direction more than a hundred years later.

$$H(\mathbf{r}, \mathbf{p}) = \sum_a \frac{\mathbf{p}_a^2}{2m_a} + U, \quad U(\mathbf{r}) = -\sum_{a<b} \frac{k_{ab}}{|\mathbf{r}_a - \mathbf{r}_b|^2}$$

has the symmetry

$$H(\lambda \mathbf{r}, \lambda^{-1}\mathbf{p}) = \lambda^{-2} H(\mathbf{r}, \mathbf{p})$$

This leads to the "almost conservation" law for the generator of this canonical transformation

$$D = \sum_a \mathbf{r}_a \cdot \mathbf{p}_a$$

$$\frac{dD}{dt} = 2H$$

Since

$$D = \frac{d}{dt} I, \quad I = \frac{1}{2} \sum_a m_a \mathbf{r}_a^2$$

we get

$$\frac{d^2}{dt^2} I = 2H$$

Consider the special case that the total angular momentum (which is conserved) is zero

$$\mathbf{L} = \sum_a \mathbf{r}_a \times \mathbf{p}_a = 0$$

The equation above for the second derivative of I has drastic consequences for the stability of the system. If $H > 0$, the moment of inertia is a convex function of time: the system will eventually expand to infinite size. If $H < 0$, we have the opposite behavior and the system will collapse to its center of mass in a finite amount of time. Thus the only stable situation is when $H = 0$.

11.5. Montgomery's pair of pants

In the case $H = \mathbf{L} = 0$, we can reduce (Montgomery, 2005) the planar three body orbits with the $\frac{1}{r^2}$ potential and equal masses to the geodesics of a metric on a four-dimensional space (i.e., two complex dimensions, if we think of $R_{1,2}$ as complex numbers $z_{1,2}$).

$$ds^2 = \left[\frac{1}{|z_1|^2} + \frac{1}{|z_2 + \frac{1}{2}z_1|^2} + \frac{1}{|z_2 - \frac{1}{2}z_1|^2} \right] \left\{ |dz_1|^2 + \frac{2}{3}|dz_2|^2 \right\}$$

There is an isometry (symmetry) $z_a \to \lambda z_a$, $0 \neq \lambda \in \mathbb{C}$, which combines rotations and scaling. We can use this to remove two dimensions to get a metric on $\mathbb{C}$

$$ds^2 = U(z)|dz|^2$$

where the effective potential

$$U(z) = \frac{7}{6} \left[\frac{1}{|z-1|^2} + \frac{1}{|z|^2} + |z|^2 \right]$$

is a positive function of

$$z = \frac{z_2 + \frac{1}{2}z_1}{z_2 - \frac{1}{2}z_1}$$

It is singular at the points $z = 0, 1, \infty$, corresponding to pairwise collisions. These singular points are at an infinite distance away in this metric. (This distance has the meaning of action. So the collisions can happen in finite time, even if the distance is infinite.) Near each singularity the metric looks asymptotically like a cylinder: rather like a pair of pants for a tall thin person. Thus, we get a metric that is complete on the Riemann sphere with three points removed. Topologically, this is the same as the plane with two points removed: $C - \{0, 1\}$.

From Riemannian geometry (Morse theory), we know that there is a minimizing geodesic in each homotopy class: there is an orbit that *minimizes* the action with a prescribed sequence of turns around each singularity. Let A be the homotopy class of curves that wind around 0 once in the counter-clockwise direction and B one that winds around 1. Then A^{-1} and B^{-1} wind around $0, 1$ in the clockwise direction. Any homotopy class of closed curves corresponds to a finite word made of these two letters

$$A^{m_1} B^{n_1} A^{m_2} B^{n_2} \cdots$$

or

$$B^{n_1} A^{m_1} A^{m_2} B^{n_2} \cdots$$

with non-zero m_a, n_a. These form a group F_2, the free group on two generators: A and B do not commute. Indeed they satisfy no relations among each other at all. There are an exponentially large number of distinct words of a given length: F_2 is a *hyperbolic group*. Given each such word we have a minimizing orbit that winds around 0 a certain number m_1 times, then around 1 a certain number n_1 times then again around 0 some number m_1 times and so on.

Moreover, the curvature of the metric is negative everywhere except at two points (Lagrange points) where it is zero. Thus the geodesics diverge from each other everywhere. A small change in the initial conditions can make the orbit career off in some unpredictable direction, with a completely different sequence of A's and B's. This is chaos. The more realistic $\frac{1}{r}$ potential is harder to analyze, but is believed to have similar qualitative behavior.

Problem 11.2: Find the formula [11.5] for the function $U(z)$ by making changes of variables

$$z_1 = (z-1)e^\lambda, \quad z_2 = \frac{z+1}{2}e^\lambda$$

in the metric [11.5]. Compute the curvature (Ricci scalar) in terms of derivatives of U. Show that the curvature is zero near the apparent singularities $z = 0, 1, \infty$ corresponding to collisions. Find local changes of variables near these points to show that the metric is that of a cylinder asymptotically. Plot some geodesics of this metric by numerical calculation.

Problem 11.3: *Virial Theorem* Show that for a bound gravitational system (i.e., the bodies orbit each other for ever) twice the average kinetic energy is equal to the negative of the average potential energy. The averages are taken over a long time along a solution:

$$< f > = \lim_{\tau \to \infty} \frac{1}{\tau} \int_0^\tau f(\mathbf{r}_a(t), \mathbf{p}_a(t)) dt$$

The key is to show that $2T + U$ is a total time derivative, and so has zero average.

Problem 11.4: *Research project**** Find the action of the minimizing geodesic for each element of F_2. Use this to evaluate Gutzwiller's trace formula for $\zeta(s)$, the sum over closed orbits. Compare with the Selberg-zeta function of Riemann surfaces.

12
The restricted three body problem

A particular case of the three body problem is of historical importance in astronomy: when one of the bodies is of infinitesimal mass m_3 and the other two bodies (the *primaries* of masses m_1, m_2) are in circular orbit around their center of mass; moreover, the orbit of the small body lies in the same plane as this circle. This is a good approximation for a satellite moving under the influence of the Earth and the Moon; an asteroid with the Sun and Jupiter; a particle in a ring of Saturn influenced also by one of its moons. The basic results are due to Lagrange, but there are refinements (e.g., "halo orbits") being discovered even in our time.

12.1. The motion of the primaries

Since the secondary has infinitesimal mass, its effect on the primaries can be ignored. Choose a reference frame where the center of mass of the primaries is at rest at the origin. The relative co-ordinate will describe an ellipse. We assume that the eccentricity of this orbit is zero, a circle centered at the origin. If R is the radius (the distance between the primaries) and Ω the angular velocity,

$$\frac{m_1 m_2}{m_1 + m_2} R\Omega^2 = \frac{Gm_1 m_2}{R^2}, \implies \Omega^2 R^3 = G(m_1 + m_2)$$

This is just Kepler's third law. The distance of the first primary from the c.m. is νR with $\nu = \frac{m_2}{m_1 + m_2}$. We can assume that $m_1 > m_2$ so that $\nu < \frac{1}{2}$. The other primary will be at a distance $(1 - \nu)R$ in the opposite direction. Thus the positions of the primaries are, in Cartesian co-ordinates,

$$(\nu R \cos \Omega t, \nu R \sin \Omega t) \quad \text{and} \quad (-[1 - \nu]R\cos\Omega t, -[1 - \nu]R\sin\Omega t)$$

12.2. The Lagrangian

The secondary will move in the gravitational field created by the two primaries. This field is time dependent, with a period equal to $\frac{2\pi}{\Omega}$. The Lagrangian is

$$L = \frac{1}{2}m_3 \dot{r}^2 + \frac{1}{2}m_3 r^2 \dot{\phi}^2 + G(m_1 + m_2)m_3 \left[\frac{1 - \nu}{\rho_1(t)} + \frac{\nu}{\rho_2(t)} \right]$$

where $\rho_{1,2}(t)$ are the distances to the primaries:

$$\rho_1(t) = \sqrt{\left[r^2 + \nu^2 R^2 - 2\nu r R \cos[\phi - \Omega t]\right]}$$
$$\rho_2(t) = \sqrt{\left[r^2 + (1-\nu)^2 R^2 + 2(1-\nu)r R \cos[\phi - \Omega t]\right]}$$

We can choose a unit of mass so that $m_3 = 1$, a unit of distance so that $R = 1$ and a unit of time so that $\Omega = 1$. Then $G(m_1 + m_2) = 1$ as well. The Lagrangian simplifies to

$$L = \frac{1}{2}\dot{r}^2 + \frac{1}{2}r^2\dot{\phi}^2 + \left[\frac{1-\nu}{\rho_1(t)} + \frac{\nu}{\rho_2(t)}\right]$$

Since the Lagrangian is time dependent, energy is not conserved: the secondary can extract energy from the rotation of the primaries. But we can make a transformation to a rotating co-ordinate

$$\chi = \phi - t$$

to eliminate this time dependence

$$L = \frac{1}{2}\dot{r}^2 + \frac{1}{2}r^2[\dot{\chi} + 1]^2 + \left[\frac{1-\nu}{r_1} + \frac{\nu}{r_2}\right]$$

where

$$r_1 = \sqrt{\left[r^2 + \nu^2 - 2\nu r \cos\chi\right]}, \quad r_2 = \sqrt{\left[r^2 + (1-\nu)^2 + 2(1-\nu)r \cos\chi\right]}$$

We pay a small price for this: there are terms in the Lagrangian that depend on $\dot{\chi}$ linearly. These lead to velocity dependent forces (the *Coriolis force*) in addition to the more familiar centrifugal force. Nevertheless, we gain a conserved quantity, the hamiltonian.

$$H = \dot{r}\frac{\partial L}{\partial \dot{r}} + \dot{\chi}\frac{\partial L}{\partial \dot{\chi}} - L$$
$$H = \frac{1}{2}\dot{r}^2 + \frac{1}{2}r^2\dot{\chi}^2 - \left[\frac{r^2}{2} + \frac{1-\nu}{r_1} + \frac{\nu}{r_2}\right]$$

This is the sum of kinetic energy and a potential energy of the secondary; it is often called the *Jacobi integral*. ("Integral" in an old term for a conserved quantity.) The Coriolis force does no work, being normal to the velocity always, so it does not contribute to the energy. It is important this is the energy measured in a non-inertial reference frame, which is why it includes the term $\propto r^2$, the centrifugal potential.

The Lagrangian can also be written in rotating Cartesian co-ordinates

$$L = \frac{1}{2}\dot{x}^2 + \frac{1}{2}\dot{y}^2 + \Omega\left[x\dot{y} - y\dot{x}\right] - V(x, y)$$
$$V(x, y) = -\left[\frac{x^2 + y^2}{2} + \frac{1-\nu}{r_1} + \frac{\nu}{r_2}\right] \tag{12.1}$$
$$r_1 = \sqrt{(x-\nu)^2 + y^2}, \quad r_2 = \sqrt{(x + [1-\nu])^2 + y^2}$$

Exercise 12.1: Show that this leads to a hamiltonian

$$H = \frac{1}{2}v_x^2 + \frac{1}{2}v_y^2 + V(x, y)$$

with the unusual Poisson brackets for the velocities:

$$\{v_x, x\} = 1 = \{v_y, y\}, \quad \{v_x, y\} = 0 = \{v_y, x\}, \quad \{v_x, v_y\} = -2\Omega \qquad (12.2)$$

The Coriolis force behaves much like a constant magnetic field normal to the plane; we will see later that a magnetic field modifies the Poisson brackets of velocities as well.

12.3. A useful identity

It is useful to express the potential energy in terms of r_1 and r_2, eliminating r. From the definition of $r_{1,2}$ we can verify that

$$[1 - \nu]r_1^2 + \nu r_2^2 - \nu(1 - \nu) = r^2$$

Thus

$$V = -\left[(1 - \nu)\left\{\frac{1}{r_1} + \frac{r_1^2}{2}\right\} + \nu\left\{\frac{1}{r_2} + \frac{r_2^2}{2}\right\}\right]$$

except for an additive constant. In studying the potential, we can use the distances r_1 and r_2 themselves as co-ordinates in the plane: the identity above shows that the potential is separable with this choice. But beware that this system breaks down along the line connecting the primaries. Along this line $r_1 + r_2 = 1$, and so they are not independent variables. Also, these variables cover only one half of the plane, the other half being obtained by reflection about the line connecting the primaries.

12.4. Equilibrium points

There are points where the forces are balanced such that the secondary can be at rest. (In the inertial frame, it will then rotate at the same rate as the primaries.)

A short exercise in calculus will show that there is a *maximum* of the potential when

$$r_1 = r_2 = 1$$

That is, when the thee bodies are located along the vertices of an equilateral triangle. There are actually two such points, on either side of the primary line. They are called *Lagrange points* L_4 and L_5. There are three more equilibrium points L_1, L_2, L_3 that lie along the primary line $y = 0$. They are not visible in terms of r_1 and r_2 because that system breaks down there. But in the Cartesian co-ordinates, it is clear by the symmetry $y \to -y$ that

$$\frac{\partial V}{\partial y} = 0, \quad \text{if } y = 0$$

When $y = 0$,

$$V = -\left[\frac{x^2}{2} + \frac{1-\nu}{|x - \nu|} + \frac{\nu}{|x + [1 - \nu]|}\right]$$

This function has three extrema, L_1, L_2, L_3. As functions of x these are maxima, but are minima along y: they are *saddle points* of V. There are no other equilibrium points.

12.5. Hill's regions

It is already clear that for a given H, only the region with $H - V > 0$ is accessible to the secondary particle. For small H, the curve $H = V$ is disconnected, with regions near m_1 and m_2 and near infinity: these are the places where the potential goes to $-\infty$. As H grows to the value of the potential at L_1 (the saddle point in between the two primaries), the two regions around the primaries touch; as H grows higher, they merge into a single region. It is only as H grows larger than the potential at $L_{4,5}$ that all of space is available for the secondary.

For example, if a particle is to move from a point near m_1 to one near m_2, the least amount of energy it needs to have is the potential at the Lagrange point in between them. The saddle point is like a mountain pass that has to be climbed to go from one deep valley to another. This has interesting implications for space travel, many of which have been explored in fiction. For example, in the imagination of many authors, Lagrange points would have strategic importance (like straits that separate continents) to be guarded by star cruisers. Belbruno, Marsden and others have scientifically sound ideas on transfer orbits of low fuel cost. It turns out that very low cost travel is possible if we allow orbits to wind around many times: it would take a few weeks to get to the moon instead of days as with the more direct routes. One is reminded of the era of slow travel around the world by sailing ships.

12.6. The second derivative of the potential

Th equilibrium points L_4, L_5, being at the maximum of a potential, would ordinarily be unstable. But an amazing fact discovered by Lagrange is that the velocity dependence of the Coriolis force can (if ν is not too close to a half) make them *stable* equilibria. Such a reversal of fortune does not happen for L_1, L_2, L_3: saddle points are not turned into stable equilibria. But it has been found recently (numerically) that there are orbits near these points, called halo orbits which do not cost much in terms of rocket fuel to maintain (see Fig. 12.1).

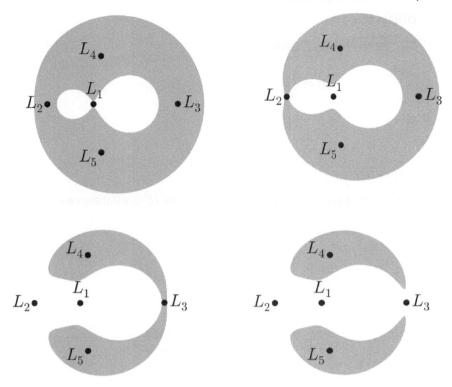

Fig. 12.1 Hill's regions for $\nu = 0.2$. The orbit cannot enter the gray region, which shrinks as H grows.

To understand the stability of L_4 and L_5 we must expand the Lagrangian to second order around them and get an equation for small perturbations. The locations of $L_{4,5}$ are

$$x = \left[\nu - \frac{1}{2}\right], \quad y = \pm\frac{\sqrt{3}}{2}$$

The second derivative of V at $L_{4,5}$ is

$$V'' \equiv K = -\begin{bmatrix} \dfrac{3}{4} & \pm\dfrac{\sqrt{27}}{4}[1 - 2\nu] \\ \pm\dfrac{\sqrt{27}}{4}[1 - 2\nu] & \dfrac{9}{4} \end{bmatrix}$$

Note that $K < 0$: both its eigenvalues are negative. On the other hand, for $L_{1,2,3}$ we will have a diagonal matrix for K with one positive eigenvalue (along y) and one negative eigenvalue (along x).

12.7.   Stability theory

The Lagrangian takes the form, calling the departure from equilibrium (q_1, q_2)

$$L \approx \frac{1}{2}\dot{q}_i\dot{q}_i - \frac{1}{2}F_{ij}q_i\dot{q}_j - \frac{1}{2}K_{ij}q_iq_j$$

where

$$F = 2\begin{bmatrix} 0 & -1 \\ 1 & 0 \end{bmatrix}$$

comes from the Coriolis force. For small q, the equations of motion become

$$\ddot{q} + F\dot{q} + Kq = 0$$

We seek solutions of the form

$$q(t) = e^{i\omega t}A$$

for some constant vector and frequency ω. Real values of ω would describe stable perturbations. The eigenvalue equation is somewhat unusual

$$[-\omega^2 + Fi\omega + K]A = 0$$

in that it involves both ω and ω^2. Thus the characteristic equation is

$$\det[-\omega^2 + Fi\omega + K] = 0$$

Or

$$\det\begin{bmatrix} K_{11} - \omega^2 & K_{12} - 2i\omega \\ K_{12} + 2i\omega & K_{22} - \omega^2 \end{bmatrix} = 0$$

which becomes

$$\omega^4 - [4 + \mathrm{tr}K]\omega^2 + \det K = 0$$

There are two roots for ω^2.

$$(\omega^2 - \lambda_1)(\omega^2 - \lambda_2) = 0$$

The condition for stability is that both roots must be real and positive. This is equivalent to requiring that the discriminant is positive and also that $\lambda_1\lambda_2 > 0, \lambda_1 + \lambda_2 > 0$. Thus

$$[\mathrm{tr}K + 4]^2 - 4\det K > 0, \quad \det K > 0, \quad \mathrm{tr}K + 4 > 0$$

So,

$$\det K > 0, \quad \operatorname{tr}K + 4 > 2\sqrt{\det K}$$

The first condition cannot be satisfied by a saddle point of V: the eigenvalues of K has opposite signs. So $L_{1,2,3}$ are unstable equilibria. It can be satisfied by a minimum, for which $K > 0$.

But surprisingly, it can also be satisfied by a maximum of a potential. For $L_{4,5}$

$$\operatorname{tr}K = -3, \quad \det K = \frac{27}{4}\nu(1-\nu)$$

Since we chose $\nu < \frac{1}{2}$, we see that $\det K > 0$.

The second condition of (12.7) becomes

$$27\nu(1-\nu) < 1$$

In other words,

$$\nu < \frac{1}{2}\left[1 - \sqrt{1 - \frac{4}{27}}\right] \approx 0.03852$$

(Recall that we chose $\nu < \frac{1}{2}$; by calling m_1 the mass of the larger primary.) Thus we get stability if the masses of the two primaries are sufficiently different from each other. In this case, the frequencies are

$$\omega = \sqrt{\frac{1 \pm \sqrt{1 - 27\nu(1-\nu)}}{2}}$$

in units of Ω.

When $\nu \ll 1$, one of these frequencies will be very small, meaning that the orbit is nearly synchronous with the primaries.

For the Sun–Jupiter system, $\nu = 9.5388 \times 10^{-4}$ so the Lagrange points are stable. The periods of libration (the small oscillations around the equilibrium) follow from the orbital period of Jupiter (11.86 years): 147.54 years or 11.9 years. For the Earth–Moon system $\nu = \frac{1}{81}$ is still small enough for stability. The orbital period being 27.32 days, we have libration periods of 90.8 days and 28.6 days.

Lagrange discovered something even more astonishing: the equilateral triangle is a stable exact solution for the *full three body problem*, not assuming one of the bodies to be infinitesimally small. He thought that these special solutions were artificial and that they would never be realized in nature. But we now know that there are asteroids (*Trojan asteroids*) that form an equilateral triangle with the Sun and Jupiter. The Earth also has such a co-traveller at its Lagrange point with the Sun.

Problem 12.2: What is the minimum energy per unit mass needed to travel from a near-Earth orbit (100 km from the Earth's surface) to the surface of the Moon?

Problem 12.3: Suppose that a particle in libration around a Lagrange point is subject to a frictional force proportional to its velocity. Will it move closer to or farther away from the Lagrange point? Find its orbit in the linear approximation in the displacement. Assume that the coefficient of friction is small.

Problem 12.4: Suppose that the orbit of the primaries have a small eccentricity ϵ. To the leading order in ϵ, find the correction to the equation of motion of the secondary.

Problem 12.5: Without assuming that the masses are very different from each other, show that there is a solution of the three body problem where the bodies are equidistant from each other.

13
Magnetic fields

The force on a charged particle in a static magnetic field is normal to its velocity. So it does no work on the particle. The total energy of the particle is not affected by the magnetic field. The hamiltonian as a function of position and velocities does not involve the magnetic field. Can Hamiltonian mechanics still be used to describe such systems? If so, where does the information about the magnetic field go in? It turns out that the magnetic field modifies the Poisson brackets and not the hamiltonian.

13.1. The equations of motion

$$\frac{d\mathbf{r}}{dt} = \mathbf{v}, \quad \frac{d}{d}[m\mathbf{v}] = e\mathbf{v} \times \mathbf{B}$$

Or in terms of components

$$m\frac{dx^i}{dt} = v^i, \quad \frac{d}{dt}[mv^i] = e\epsilon_{ijk}v^j B^k$$

Here ϵ_{ijk} is completely antisymmetric and

$$\epsilon_{123} = 1$$

Let us assume that the magnetic field does not depend on time, only on the position; otherwise, we cannot ignore the electric field.

13.2. Hamiltonian formalism

The energy (hamiltonian) is just

$$H = \frac{1}{2}mv^i v^i$$

We want however,

$$\{H, x^i\} = v^i, \quad \{H, v_i\} = \frac{e}{m}\epsilon_{ijk}v^j B^k$$

The first is satisfied if

$$\{p^i, x^j\} = \frac{1}{m}\delta^{ij}$$

which is the usual relation following from canonical relations between position and momentum. So we want

$$\left\{\frac{1}{2}mv^j v^j, v^i\right\} = \frac{e}{m}\epsilon_{ijk}v^j B^k$$

Using the Leibnitz rule this becomes

$$v^j \{v^j, v_i\} = \frac{e}{m^2}\epsilon_{ijk}v^j B^k$$

It is therefore sufficient that

$$\{v_j, v_i\} = \frac{e}{m^2}\epsilon_{ijk}B^k$$

This is *not* what follows from canonical relations; the different components of momentum would commute then. To make this distinction clear, let us denote momentum by

$$\pi_i = mv_i$$

Then:

$$\{x^i, x^j\} = 0, \quad \{\pi^i, x^j\} = \delta^{ij}, \quad \{\pi_i, \pi_i\} = -eF_{ij}$$

where

$$F_{ij} = \epsilon_{ijk}B_k$$

The brackets are antisymmetric. The Jacobi identity is automatic for all triples of variables except one:

$$\{\{\pi_i, \pi_j\}, \pi_k\} = -e\{F_{ij}, mv_k\}$$
$$= e\partial_k F_{ij}$$

Taking the cyclic sum, we get

$$\partial_i F_{jk} + \partial_j F_{ki} + \partial_k F_{ij} = 0$$

If you work out in components you will see that this is the condition

$$\nabla \cdot \mathbf{B} = 0$$

which is one of Maxwell's equations. Thus the Jacobi identity is satisfied as long as Maxwell's equation is satisfied.

If we have an electrostatic as well as a magnetic field the hamiltonian will be

$$H = \frac{\pi_i \pi_i}{2m} + eV$$

again with the commutation relations above.

13.3.   Canonical momentum

It is possible to bring the commutation relations back to the standard form

$$\{x^i, x^j\} = 0, \quad \{p^i, x^j\} = \delta^{ij}, \quad \{p_i, p_i\} = 0$$

in those cases where the magnetic field is a curl. Recall that locally, every field satisfying

$$\nabla \cdot \mathbf{B} = 0$$

is of the form

$$\mathbf{B} = \nabla \times \mathbf{A}$$

for some vector field $\mathbf{A}$. This is not unique: a change (*gauge transformation*)

$$\mathbf{A} \to \mathbf{A} + \nabla \Lambda$$

leaves $\mathbf{B}$ unchanged. Now if we define

$$\pi_i = p_i - eA_i$$

then the canonical relations imply the relations for π_i.

13.4.   The Lagrangian

This suggests that we can find a Lagrangian in terms of A_i. We need

$$p_i = \frac{\partial L}{\partial \dot{x}^i}$$

or

$$\frac{\partial L}{\partial \dot{x}^i} = m\dot{x}_i + eA_i$$

Thus we propose

$$L = \frac{1}{2}m\dot{x}^i\dot{x}^i + eA_i\dot{x}^i - eV$$

as the Lagrangian for a particle in an electromagnetic field.

Exercise 13.1: Show that the Lorentz force equations follows from this Lagrangian.

An important principle of electromagnetism is that the equations of motion should be invariant under gauge transformations $A_i \to A_i + \partial_i\Lambda$. Under this change the action changes to

$$S = \int_{t_1}^{t_2} L\,dt \to S + \int_{t_1}^{t_2} e\dot{x}^i\partial_i\Lambda\,dt$$

The extra term is a total derivative, hence only depends on endpoints:

$$\int_{t_1}^{t_2} e\frac{d\Lambda}{dt}\,dt = e\left[\Lambda(x(t_2)) - \Lambda(x(t_1))\right]$$

Since we hold the endpoints fixed during a variation, this will not affect the equations of motion.

13.5. The magnetic monopole*

Recall that the electric field of a point particle satisfies

$$\nabla \cdot \mathbf{E} = 0$$

everywhere but at its location. Can there be point particles that can serve as sources of magnetic fields in the same way? None have been discovered to date: only magnetic dipoles have been found, a combination of north and south poles. Dirac discovered that the existence of even one such magnetic monopole somewhere in the universe would explain a remarkable fact about nature: that electric charges appear as multiples of a fundamental unit of charge. To understand this let us study the dynamics of an electrically charged particle in the field of a magnetic monopole, an analysis due to M. N. Saha.

$$\frac{d\mathbf{r}}{dt} = \mathbf{v}, \quad \frac{d}{d}[m\mathbf{v}] = eg\mathbf{v} \times \frac{\mathbf{r}}{r^3}$$

where r is the strength of the magnetic monopole. The problem has spherical symmetry, so we should expect angular momentum to be conserved. But we can check that

$$\frac{d}{dt}[\mathbf{r} \times m\mathbf{v}] = eg\mathbf{r} \times \left[\mathbf{v} \times \frac{\mathbf{r}}{r^3}\right]$$

is not zero. What is going on? Now, recall that identity

$$\frac{d}{dt}\left[\frac{\mathbf{r}}{r}\right] = \frac{\mathbf{v}}{r} - \frac{\mathbf{r}}{r^2}\dot{r}$$

But

$$\dot{r} = \frac{1}{2r}\frac{d}{dt}[\mathbf{r}\cdot\mathbf{r}]$$

$$= \frac{1}{r}\mathbf{r}\cdot\mathbf{v}$$

So

$$\frac{d}{dt}\left[\frac{\mathbf{r}}{r}\right] = \frac{r^2\mathbf{v} - \mathbf{r}(\mathbf{v}\cdot\mathbf{r})}{r^3}$$

or

$$\frac{d}{dt}\left[\frac{\mathbf{r}}{r}\right] = \frac{1}{r^3}\mathbf{r}\times[\mathbf{v}\times\mathbf{r}]$$

Thus we get a new conservation law

$$\frac{d}{dt}\left[\mathbf{r}\times m\mathbf{v} - eg\frac{\mathbf{r}}{r}\right] = 0$$

The conserved angular momentum is the sum of the orbital angular momentum and a vector pointed along the line connecting the charge and the monopole. This can be understood as the angular momentum contained in the electromagnetic field. When an electric and a magnetic field exist together, they carry not only energy but also momentum and angular momentum. If you integrate the angular momentum density over all of space in the situation above, you will get exactly the extra term

$$\mathbf{J} = \mathbf{r}\times m\mathbf{v} - eg\frac{\mathbf{r}}{r}$$

which is a fixed vector in space. The orbit does not lie in the plane normal to to $\mathbf{J}$. Instead it lies on a cone, whose axis is along $\mathbf{J}$. The angle α of this cone is given by

$$J\cos\alpha = \mathbf{J}\cdot\frac{\mathbf{r}}{r} = -eg$$

13.5.1   Quantization of electric charge

If we quantize this system, we know that an eigenvalue of J_3 is an integer or half-integer multiple of $\hbar$ and that the eigenvalues of J^2 are $j(j+1)$ where j is also such a multiple. On the other hand, $\mathbf{L} = \mathbf{r}\times m\mathbf{v}$ is also quantized in the same way. It follows that the vector $eg\frac{\mathbf{r}}{r}$ must have a magnitude which is a multiple of $\hbar$

$$eg = n\hbar, \quad n = \frac{1}{2}, 1, \frac{3}{2}, 2, \cdots$$

Thus, if there is even one magnetic monopole somewhere in the universe, electric charge has to be quantized in multiples of $\frac{\hbar}{2g}$. We do see that electric charge is quantized this way, the basic unit being the magnitude of the charge of the electron. We do not yet know if the reason for this is the existence of a magnetic monopole.

13.6.   The Penning trap

A static electromagnetic field can be used to trap charged particles. You can bottle up anti-matter this way; or use it to hold an electron or ion in place to make precise measurements on it.

It is not possible for a static electric field by itself to provide a stable equilibrium point: the potential must satisfy the Laplace equation $\nabla^2 V = 0$. So at any point the sum of the eigenvalues of the Hessian matrix V'' (second derivatives) must vanish. At a stable minimum they would all have to be positive. It is possible to have one negative and two positive eigenvalues. This is true for example for a quadrupole field

$$V(x) = \frac{k}{2} \left[2x_3^2 - x_1^2 - x_2^2 \right]$$

Such a field can be created by using two electrically charged conducting plates shaped like the sheets of a hyperboloid:

$$x_1^2 + x_2^2 - 2x_3^2 = \text{constant}$$

So, the motion in the $x_1 - x_2$ plane is unstable but that along the x_3 axis is stable. Now we can put a constant magnetic field pointed along the x_3-axis. If it is strong enough, we get a stable equilibrium point at the origin. This stabilization of a maximum by velocity-dependent forces is the phenomenon we saw first at the Lagrange points of the three body problem.

Look at the equation of motion of a particle in a constant magnetic field and an electrostatic potential that is a quadratic function of position, $V(x) = \frac{1}{2} x^T K x$.

$$\ddot{q} + F\dot{q} + \frac{e}{m} K q = 0$$

Here F is an antisymmetric matrix proportional to the magnetic field. For a field of magnitude B along the x_3-axis

$$F = \frac{e}{m} \begin{pmatrix} 0 & B & 0 \\ -B & 0 & 0 \\ 0 & 0 & 0 \end{pmatrix}$$

If we assume the ansatz

$$q = A e^{i\omega t}$$

the equation becomes

$$\left[-\omega^2 + i\omega F + \frac{e}{m} K \right] A = 0$$

which gives the characteristic equation

$$\det \left[-\omega^2 + i\omega F + \frac{e}{m} K \right] = 0$$

If the magnetic field is along the third axis and if the other matrix has the form

$$K = \begin{pmatrix} k_1 & 0 & 0 \\ 0 & k_2 & 0 \\ 0 & 0 & k_3 \end{pmatrix}$$

this equation factorizes

$$(-\omega^2 + k_3) \det \begin{bmatrix} \dfrac{e}{m} k_1 - \omega^2 & i\omega \dfrac{eB}{m} \\ -i\omega \dfrac{eB}{m} & \dfrac{e}{m} k_2 - \omega^2 \end{bmatrix} = 0$$

The condition for stability is that all roots for ω^2 are positive. One of the roots is k_3, so it must be positive. It is now enough that the discriminant as well as the sum and product of the other two roots are positive. This amounts to

$$k_3 > 0, \quad k_1 k_2 > 0, \quad \frac{eB^2}{m} > 2\sqrt{k_1 k_2} - (k_1 + k_2)$$

The physical meaning is that the electrostatic potential stabilizes the motion in the x_3-direction. Although the electric field pushes the particle away from the origin, the magnetic force pushes it back in.

A collection of particles moving in such a trap will have frequencies dependent on the ratios $\frac{e}{m}$. These can be measured by Fourier transforming the electric current they induce on a probe. Thus we can measure the ratios $\frac{e}{m}$, a technique called *Fourier transform mass spectrometry*.

Problem 13.2: Find the libration frequencies of a charged particle in a Penning trap.

Problem 13.3: Verify that $\mathbf{J}$ of a charged particle in the field of a magnetic monopole satisfies the commutation relations of angular momentum

$$\{J_i, J_j\} = \epsilon_{ijk} J_k$$

Does orbital angular momentum $\mathbf{r} \times m\mathbf{v}$ satisfy these relations?

Problem 13.4: Determine the orbit of a charged particle moving in the field of a fixed magnetic monopole. Exploit the conservation of energy and angular momentum. You should find that the orbit lies on a cone whose axis is along $\mathbf{J}$.

14
Poisson and symplectic manifolds

Classical mechanics is not just old mechanics. It is the approximation to mechanics in which the quantum effects are small. By taking limits of quantum systems, we sometimes discover new classical systems that were unknown to Hamilton or Euler. A good example of this is the concept of spin. It used to be thought that the only way a particle can have angular momentum is if it moves: $\mathbf{L} = \mathbf{r} \times m\dot{\mathbf{r}}$. But we saw a counter-example in the last chapter: a charge and a magnetic monopole can have angular momentum even if they are at rest, because the electric and magnetic fields carry angular momentum. But this is an exotic example, one that has not yet been realized experimentally.

But almost every particle (the electron, the proton, the neutron etc.) of which ordinary matter is made of carries *spin*: it has angular momentum even when it is at rest. For each particle this is small (of order $\hbar$), so it is outside of the domain of classical physics. But a large number of such particles can act in unison to create a spin big enough to be treated classically. An example of this is nuclear magnetic resonance, where a large number ($N \approx 10^{23}$) of atomic nuclei at rest orient their spins in the same direction. This happens because they carry a magnetic moment parallel to the spin, and if you apply a magnetic field, the minimum energy configuration is to have the spins line up. The time evolution of this spin can be described classically, as the spin is large compared to $\hbar$.

There are some unusual things about this system though: its phase space has finite area. It is a sphere. So it will not be possible to separate its co-ordinates into conjugate variables globally. This does not matter: we still have a Poisson bracket and can still derive a hamiltonian formalism.

14.1. Poisson brackets on the sphere

The essential symmetry of the sphere is under rotations. Recall the commutation relations of infinitesimal rotations

$$\{S_1, S_2\} = S_3, \quad \{S_2, S_3\} = S_1, \quad \{S_3, S_1\} = S_1$$

Or, in the index notation

$$\{S_i, S_j\} = \epsilon_{ijk} S_k$$

In particular, the Jacobi identity is satisfied. Note that $\mathbf{S} \cdot \mathbf{S}$ commutes with every component. So we can impose the condition that it is a constant

$$S^2 = \left\{ \mathbf{S} | \mathbf{S} \cdot \mathbf{S} = s^2 \right\}$$

This way, we get a Poisson bracket of a pair of functions on the sphere

$$\{F, G\} = \epsilon_{ijk} S_i \frac{\partial F}{\partial S_j} \times \frac{\partial G}{\partial S_k}$$

Or equivalently,

$$\{F, G\} = \mathbf{S} \cdot \frac{\partial F}{\partial \mathbf{S}} \times \frac{\partial G}{\partial \mathbf{S}}$$

If you find this formulation too abstract, you can use spherical polar co-ordinates

$$S_1 = s \sin\theta \cos\phi, \quad S_2 = s \sin\theta \sin\phi, \quad s_3 = s \cos\theta$$

so that the Poisson bracket can be written as

$$\{\phi, \cos\theta\} = 1$$

In some sense ϕ and $\cos\theta$ are canonically conjugate variables. But this interpretation is not entirely right: unlike the usual canonical variables, these are bounded. Also, they are not globally well-defined co-ordinates on the sphere. For example, ϕ jumps by 2π when we make a complete rotation. This won't matter as long as only local properties in the neighborhood of some point are studied. But global questions (such as quantization or chaos) are best studied in a co-ordinate independent way.

14.2.   Equations of motion

Consider a large number of identical particles, each with a magnetic moment proportional to its spin. Then the total magnetic moment will also be proportional to spin:

$$\mathbf{M} = g\mathbf{S}$$

In the presence of an external magnetic field the energy of this system will be

$$H = g\mathbf{B} \cdot \mathbf{S}$$

The time evolution of the spin is then given by Hamilton's equations

$$\frac{d\mathbf{S}}{dt} = g\mathbf{B} \times \mathbf{S}$$

The solution is easy to find: the spin precesses at a constant rate around the magnetic field. The frequency of this precession is

$$\omega = g|\mathbf{B}|$$

If the initial direction of the magnetic moment was parallel to the magnetic field (minimum energy) or antiparallel (maximum of energy) the spin will not precess. Otherwise it will precess in a circle determined by the energy. Suppose that all the magnetic moments are in the lowest energy state, aligned with the magnetic field. If we now send in an electromagnetic wave at just this frequency, the spins will absorb the energy and start to precess. This is the phenomenon of nuclear magnetic resonance (NMR). This absorption can be detected and used to measure the density of the spins. If we create a slowly varying magnetic field (instead of a constant one as we have assumed so far) the resonant frequency will depend on position. By measuring the amount of electromagnetic energy absorbed at each frequency, we can deduce the density of nuclei at each position. This is the technique of NMR imaging.

14.3. Poisson manifolds

Thus we see that hamiltonian mechanics makes sense even when it is not possible to find global canonical co-ordinates. All we need is a Poisson bracket among functions that satisfies the following axioms:

$$\{F, aG + bH\} = a\{F, G\} + b\{F, H\}, \quad a, b \in R, \quad \text{linearity}$$
$$\{F, GH\} = \{F, G\}H + F\{G, H\}, \quad \text{Leibnitz identity}$$
$$\{F, G\} = -\{G, F\}, \quad \text{antisymmetry}$$
$$\{\{F, G\}, H\} + \{\{G, H\}, F\} + \{\{H, F\}, G\} = 0, \quad \text{Jacobi identity}$$

The first two say that there is a second rank contra-variant tensor r such that

$$\{F, G\} = r^{ij}\partial_i F \partial_j G$$

This tensor must be antisymmetric

$$r^{ij} = -r^{ji}$$

The Jacobi identity is then a condition on its derivatives

$$r^{im}\partial_m r^{jk} + r^{jm}\partial_m r^{ki} + r^{km}\partial_m r^{ij} = 0$$

A manifold along with such an tensor on it is called a *Poisson manifold*. The sphere is an example, as saw above.

Example 14.1: Three-dimensional Euclidean space R^3 is a Poisson manifold with $r^{ij} = \epsilon^{ijk} x_k$. Verify the Jacobi identity.

A particularly interesting kind of Poisson manifold is an *irreducible* one: the only central functions (i.e., those that commute with all other functions) are constants:

$$\{F, G\} = 0, \forall G \implies \partial_i F = 0$$

R^3 is not irreducible: $|\mathbf{x}|^2 = x_i x_i$ commutes with everything. By setting it equal to a constant, we get a sphere which is irreducible. More generally, a Poisson manifold can be (locally) decomposed as a union of irreducibles, which are the submanifolds on which the central functions are held constant.

On an irreducible Poisson manifold, r^{ij} has an inverse

$$r^{ij} \omega_{jk} = \delta^i_k$$

In terms of this inverse, the Jacobi identity is a linear equation:

$$\partial_i \omega_{jk} + \partial_j \omega_{ki} + \partial_k \omega_{ij} = 0$$

This condition has an elegant meaning in differential geometry: $\omega = \omega_{ij} dx^i \wedge dx^j$ is a closed differential form, $d\omega = 0$. A closed invertible differential form is called a *symplectic form*; thus irreducible Poisson manifolds are symplectic manifolds. The sphere is the simplest example.

Exercise 14.1: Show that all symplectic manifolds are even-dimensional. (Can an antisymmetric matrix of odd dimension have non-zero determinant?)

There are local co-ordinates (q, p) on a symplectic manifold such that ω is a constant,

$$\omega = dq^i \wedge dp_i$$

In these coordinates the Poisson bracket reduces to the form

$$\{F, G\} = \sum_i \left(\frac{\partial F}{\partial p_i} \frac{\partial G}{\partial q^i} - \frac{\partial F}{\partial q^i} \frac{\partial G}{\partial p_i} \right)$$

we had earlier. But the co-ordinate independent point of view is more natural and allows for the treatment of many more physical systems (such as spin).

14.4.   Liouville's theorem

The determinant of an antisymmetric matrix of even dimension is the square of a polynomial in its matrix elements. This polynomial is called the Pfaffian. The Pfaffian of the symplectic form is density in the phase space. As canonical transformations leave the symplectic form unchanged, they leave this density unchanged as well. In the simplest case $\omega = dq^i \wedge dp_i$ the density is just $dqdp$. That the volume element in phase space, $dqdp$, is invariant under time evolution is an ancient result, known as Liouville's theorem.

Problem 14.2: In a ferromagnetic material, each atom has an unpaired electron which carries a magnetic moment and $\frac{\hbar}{2}$ of spin. What is the spin of a ferromagnet consisting of one kilogram of iron, in which about a tenth of the unpaired electrons are lined up?

Problem 14.3: A Grassmannian is a generalization of the sphere. It may be thought of as the set of all $M \times M$ hermitian matrices equal to their own squares (orthogonal projectors). Such matrices have eigenvalues 1 or 0. The number of non-zero eigenvalues is called the rank.

$$\mathrm{Gr}_m(M) = \{P | P^2 = P, P^\dagger = P, \mathrm{tr}\ P = m\}$$

Show that: The case $M = 2$, $m = 1$ corresponds to the sphere. $\omega = \mathrm{tr}\ PdP \wedge dP$ is a symplectic form on the $\mathrm{Gr}_m(M)$. The Poisson brackets of the matrix elements of P are

$$\{P^i_j, P^k_l\} = \delta^k_j P^i_l - \delta^i_l P^k_j$$

15
Discrete time

Is the flow of time continuous? Or is it discrete, like sand in an hourglass? As far as we know now, it is continuous. Yet, it is convenient in many situations to think of it as discrete. For example, in solving a differential equation numerically, it is convenient to calculate the finite change over a small time interval and iterate this process. It is important that we retain the symmetries and conservation laws of the differential equation in making this discrete approximation. In particular, each time step must be a canonical (also called symplectic) transformation. Such symplectic integrators are commonly used to solve problems in celestial mechanics.

Another reason to study discrete time evolution is more conceptual. It goes back to Poincaré's pioneering study of chaos. Suppose that we have a system with several degrees of freedom. We can try to get a partial understanding by projecting to one degree of freedom (say (p_1, q_1)). That is, we look only at initial conditions where the degrees of freedom are fixed at some values (say $(p_i, q_i) = 0$ for $i > 1$). Then we let the system evolve in the full phase space and ask when it will return to this subspace (i.e., what is the next value of (p_1, q_1) for which $(p_i, q_i) = 0$ for $i > 1$?). This gives a canonical transformation (called the *Poincaré map*) of the plane (p_1, q_1) to itself. We can then iterate this map to get an orbit, an infinite sequence of points in the plane. In simple cases (like the harmonic oscillator), this orbit will be periodic. If the system is chaotic, the orbit will wander all over the plane, and it is interesting to ask for the density of its distribution: how many points in the orbit are there in some given area? This distribution often has an interesting fractal structure: there are islands that contain points in the orbit surrounded by regions that do not contain any. But if we were to magnify these islands, we will see that they contain other islands surrounded by empty regions and so on to the smallest scales.

15.1.   First order symplectic integrators

Symplectic transformation is just another name for a canonical transformation, i.e., a transformation that preserves the Poisson brackets. In many cases of physical interest, the hamiltonian of a mechanical system is of the form

$$H = A + B$$

where the dynamics of A and B are separately are easy to solve. For example, if the hamiltonian is

$$H = T(p) + V(q)$$

with a kinetic energy T that depends on momentum variables alone and a potential energy V that depends on position alone, it is easy to solve the equations for each separately. The problem is, of course, that these two canonical transformations do not commute. So we cannot solve the dynamics of H by combining them.

But the commutator of these transformations is of order $t_1 t_2$. Thus for small time intervals it is small, and it might be a good approximation to ignore the lack of commutativity. This gives a first order approximation. If we split the time into small enough intervals, the iteration of this naive approximation might be good enough. We then iterate this time evolution to get an approximation to the orbit. Later we will see how to improve on this by including the effects of the commutator to the next order.

If we perform the above two canonical transformations consecutively, choosing equal time steps ϵ, we get the discrete time evolution

$$p_i' = p_i - \epsilon V_i(q), \quad q^{i\prime} = q^i + \epsilon T^i(p')$$

where $V_i = \frac{\partial V}{\partial q^i}$, $T^i = \frac{\partial T}{\partial p_i}$. We transform p first then q because usually it leads to simpler formulas. (See the example.) In many cases this simple "first order symplectic integrator" already gives a good numerical integration scheme.

Example 15.1: Consider the example of the simple pendulum $T = \frac{p^2}{2}$, $V = \omega^2 [1 - \cos q]$. If we perform the above two canonical transformations consecutively, choosing equal time steps ϵ, we get the discrete time evolution

$$p' = p - \epsilon \omega^2 \sin q, \quad q' = q + \epsilon p'$$

Equivalently,

$$p' = p - \epsilon \omega^2 \sin q, \quad q' = q + \epsilon p - \epsilon^2 \omega^2 \sin q$$

This is a canonical transformation:

$$\det \begin{bmatrix} \frac{\partial p'}{\partial p} & \frac{\partial p'}{\partial q} \\ \frac{\partial q'}{\partial p} & \frac{\partial q'}{\partial q} \end{bmatrix} = \det \begin{bmatrix} 1 & -\epsilon \omega^2 \cos q \\ \epsilon & 1 - \epsilon^2 \omega^2 \cos q \end{bmatrix} = 1$$

Note that this determinant is one exactly, not just to second order in ϵ.

Iterating this map we get a discrete approximation to the time evolution of the pendulum. It gives a nice periodic orbit as we expect for the pendulum (see Fig. 15.1).

Notice that if we had made another discrete approximation (which attempts to do the T and V transformations together) we would not have obtained a canonical transformation:

$$p' = p - \epsilon \omega^2 \sin q, \quad q' = q + \epsilon p$$

$$\det \begin{bmatrix} \frac{\partial p'}{\partial p} & \frac{\partial p'}{\partial q} \\ \frac{\partial q'}{\partial p} & \frac{\partial q'}{\partial q} \end{bmatrix} = \det \begin{bmatrix} 1 & -\epsilon \omega^2 \cos q \\ \epsilon & 1 \end{bmatrix} \neq 1$$

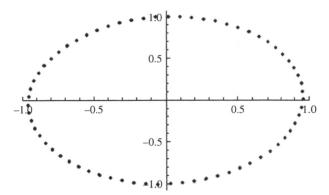

Fig. 15.1 Symplectic integration of the pendulum.

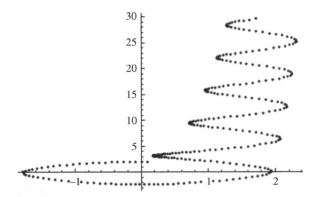

Fig. 15.2 Non-symplectic integration of the pendulum.

The orbit of this map goes wild, not respecting conservation of energy or of area (see Fig. 15.2).

15.2. Second order symplectic integrator

Suppose we are solving a linear system of differential equations

$$\frac{dx}{dt} = Ax$$

for some constant matrix A. The solution is

$$x(t) = e^{tA} x_0$$

where the exponential of a matrix is defined by the series

$$e^{tA} = 1 + tA + \frac{1}{2!} t^2 A^2 + \frac{1}{3!} t^3 A^3 + \cdots$$

Solving non-linear differential equations

$$\frac{dx^i}{dt} = A^i(x), \quad x^i(0) = x_0^i$$

is the same idea, except that the matrix is replaced by a vector field whose components can depend on x. The solution can be thought of still as an exponential of the vector field, defined by a similar power series. For example, the change of a function under time evolution has the Taylor series expansion

$$f(x(t)) = f(x_0) + tAf(x_0) + \frac{1}{2!}t^2 A^2 f(x_0) + \frac{1}{3!}t^3 A^3 f(x_0)$$

Here

$$Af = A^i \frac{\partial f}{\partial x^i}, \quad A^2 f = A^i \frac{\partial}{\partial x^i} \left(A^j \frac{\partial f}{\partial x^j} \right), \cdots$$

Now, just as for matrices,

$$e^{A+B} \neq e^A e^B$$

in general, because A and B may not commute. Up to second order in t we can ignore this effect

$$e^{t(A+B)} = e^{tA} e^{tB} \left[1 + O(t^2) \right]$$

The first order symplectic integrator we described earlier is based on this. There is a way to correct for the commutator, called the Baker–Campbell–Hausdorff formula (or Poincaré–Birkhoff–Witt lemma). To second order, it gives

Lemma 15.1:

$$e^{tA} e^{tB} = \exp \left\{ tA + tB + \frac{t^2}{2}[A, B] + O(t^3) \right\}$$

Proof: Expand the LHS to second order

$$\left[1 + tA + \frac{1}{2}t^2 A^2 \right] \left[1 + tB + \frac{1}{2}t^2 B^2 \right] = 1 + tA + tB + \frac{1}{2}t^2(A^2 + B^2) + t^2 AB + O(t^3)$$

The RHS to the second order is

$$1 + tA + tB + \frac{t^2}{2}[A, B] + \frac{1}{2}t^2(A + B)^2 + O(t^3)$$

They agree because

$$A^2 + B^2 + 2AB = (A + B)^2 + [A, B]$$

It is also possible to determine higher order corrections in the same way. $\qquad \square$

Proposition 15.1: We can avoid commutators by a more symmetric factorization

$$e^{t(A+B)} = e^{\frac{1}{2}tA}e^{tB}e^{\frac{1}{2}tA}\left[1 + \mathrm{O}(t^3)\right]$$

Proof:

$$e^{\frac{1}{2}tA}e^{tB}e^{\frac{1}{2}tA} = e^{\frac{1}{2}tA}e^{\frac{t}{2}B}\ e^{\frac{t}{2}B}e^{\frac{1}{2}tA}$$

and use the BCH formula to combine the first pair and the second pair:

$$e^{\frac{1}{2}tA}e^{tB}e^{\frac{1}{2}tA} = e^{\frac{t}{2}A+\frac{t}{2}B+\frac{t^2}{8}[A,B]+\cdots}e^{\frac{t}{2}A+\frac{t}{2}B-\frac{t^2}{8}[A,B]+\cdots}$$

and then again combine the last two exponentials. All the second order terms in t cancel out. $\qquad\qquad\square$

In our case, A and B will be generators of canonical transformations whose Hamilton's equations are easy to solve (i.e., e^{tA} and e^{tB} are known). We can then find an approximation for the solution of Hamilton's equations for $A + B$. As an example, if

$$H(q,p) = T(p) + V(q)$$

we can deduce a second order symplectic integrator (choosing $A = V, B = T$). The effect of $e^{\frac{\epsilon}{2}V}$ is to leave q unchanged and to transform p to an intermediate variable

$$z_i = p_i - \frac{1}{2}\epsilon V_i(q) \tag{15.1}$$

where $V_i = \frac{\partial V}{\partial q^i}$. Then we apply $e^{\epsilon T}$, which leaves p unchanged and transforms q to

$$q^{i\prime} = q^i + \epsilon T^i(z) \tag{15.2}$$

with $T^i = \frac{\partial T}{\partial p_i}$. Finally we apply $e^{\frac{\epsilon}{2}V}$ again to get

$$p_i' = z_i - \frac{1}{2}\epsilon V_i(q^{i\prime}) \tag{15.3}$$

The formulas (15.1, 15.2, 15.3) together implement a second order symplectic integrator. See Yoshida (1990) for higher orders.

15.3.   Chaos with one degree of freedom

A Hamiltonian system with a time independent hamiltonian always has at least one conserved quantity: the hamiltonian itself. Thus we expect all such systems with one degree of freedom to be integrable. But if the hamiltonian is time dependent this is not the case anymore. For example, if the hamiltonian is a periodic function of time, in each period the system will evolve by a canonical transformation. Iterating this we will get an orbit of a symplectic map which can often be chaotic. G. D. Birkhoff (1922) made a pioneering study of this situation.

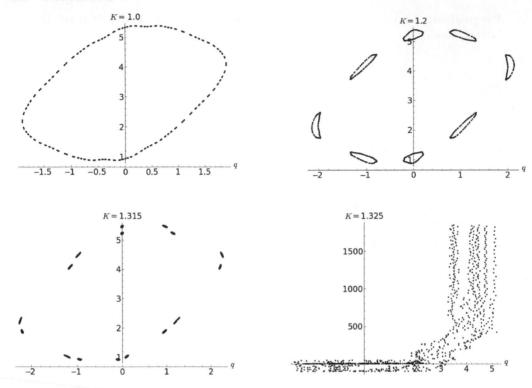

Fig. 15.3 The Chirikov standard map for the initial point $p = 0.1, q = 1$ and various values of K.

Example 15.2: The Chirikov standard map is

$$p_{n+1} = p_n + K \sin q_n, \quad q_{n+1} = q_n + p_{n+1}$$

This can be thought of as a discrete approximation to the pendulum when K is small (see the previous example): we choose units of time so that each step is one $\epsilon = 1$ and put $K = -\omega^2$.

But for K that is not small, it is an interesting dynamical system in its own right, iterating a canonical transformation. Although the evolution of the pendulum itself is integrable, this discrete evolution can be chaotic when K is not small. Look at the plots of the orbits of the same initial point $(p, q) = (0.1, 1)$ for various values of K. Thus we see that iterations of a canonical transformation can have quite unpredictable behavior even with just one degree of freedom.

Birkhoff's idea in the above-mentioned paper is to attempt to construct a conserved quantity (hamiltonian). If the time variable were continuous this would be easy. He does an analytic continuation of the discrete variable n and works backward to get a hamiltonian. This continuation is done using an asymptotic series. So the hamiltonian obtained this way is also an asymptotic series. If this series were to converge, the discrete evolution would not be chaotic. But for most initial conditions the series is only asymptotic: the lack of convergence is due to resonances. So we will find that there are regions in the phase space where the system behaves chaotically, interlaced with other regions where it is predictable. A far-reaching extension of this idea was developed by Kolmogorov, Arnold and Moser (KAM). This KAM theorem is the foundation of the modern theory of chaos in Hamiltonian systems.

Problem 15.1: Plot the orbits of the non-symplectic integrator of the pendulum

$$p' = p - \epsilon \omega^2 \sin q, \quad q' = q + \epsilon p$$

Compare with the first order symplectic integrator

$$p' = p - \epsilon \omega^2 \sin q, \quad q' = q + \epsilon p'$$

Problem 15.2: Write a program (in MatLab, Mathematica, Maple or Sage) that iterates the Chirikov standard map. Use it to plot the orbit for various values of K.

Problem 15.3: Write a symplectic integrator for the circular restricted three body problem. It is best to use $H_0 = \frac{1}{2}(v_x^2 + v_y^2)$ with Poisson brackets as in (12.2) and potential V as in (12.1). Plot some orbits and see that they lie within the corresponding Hill's region.

Problem 15.4: *Arnold's cat map* Even the iteration of a linear canonical transformation can lead to chaos, if the phase space has a periodicity that is incommensurate with it. An example is the map

$$\begin{pmatrix} x \\ y \end{pmatrix} = \begin{pmatrix} 2 & 1 \\ 1 & 1 \end{pmatrix} \begin{pmatrix} x \\ y \end{pmatrix} \mod 1$$

Verify that this is indeed a canonical transformation, i.e., that the Jacobian is unity. The origin is a fixed point. What are the stable[1] and unstable directions? When is the orbit of a point periodic? Is there an orbit that is dense? Is the map ergodic? Arnold illustrated the chaos by showing what happens to the picture of a cat under repeated applications of this map.

---

[1]Some of these terms are explained in later chapters. Come back to this project after you have read them.

16
Dynamics in one real variable

It became clear towards the end of the nineteenth century that most systems are not integrable: we will never get a solution in terms of simple functions (trigonometric, elliptic, Painlevé etc.) The focus now shifts to studying statistical properties, such as averages over long time. And to universal properties, which are independent of the details of the dynamics. It is useful to study the simplest case of dynamics, the iterations of a function of a single real variable, which maps the interval $[0, 1]$ to itself. Even a simple function like $\mu x(1 - x)$ (for a constant μ) leads to chaotic behavior. Only since the advent of digital computers has it become possible to understand this in some detail. But it is not the people who had the biggest or fastest computers that made the most important advances: using a hand-held calculator to do numerical experiments pointed Feigenbaum to patterns which led to a beautiful general theory. The best computer is still the one between your ears.

See Strogarz (2001) for more.

16.1. Maps

A map is just another word for a function $f : X \to X$ that takes some set to itself. Since the domain and range are the same, we can iterate this: given an initial point $x_0 \in X$ we can define an infinite sequence

$$x_0, \ x_1 = f(x_0), \ x_2 = f(x_1), \cdots$$

i.e.,

$$x_{n+1} = f(x_n)$$

This is an *orbit* of f. A *fixed point* of f is a point that is mapped to itself:

$$f(x) = x.$$

The orbit of a fixed point is really boring $x, x, x, \cdots$.

A *periodic orbit* satisfies

$$x_{n+k} = x_n$$

for some k. The smallest number for which this is true is called its period. For example, if the orbit of some point is periodic with period two, it will look like

$$x_0, x_1, x_0, x_1, x_0, x_1, \cdots, \quad x_0 \neq x_1$$

Given a function we can define its iterates

$$f_2(x) = f(f(x)), \quad f_3(x) = f(f(f(x))), \cdots$$

A moment's thought will show that a fixed point of one of the iterates $f_k(x)$ is the same thing as a periodic orbit of $f(x)$. For example, if x_0 is not a fixed point of $f(x)$ but is one for f_2, then its orbit under f is periodic with period two:

$$f(x_0) = x_1 \neq x_0, \quad x_0 = f(f((x_0)))$$

gives the orbit under f:

$$x_0, x_1, x_0, x_1, \cdots$$

So far we have not assumed anything about the space X or the function f. It will often be useful to put some additional conditions such as

- X is a topological space (which allows us to talk of continuous functions) and $f : X \to X$ is a continuous function
- X is a differentiable manifold and $f : X \to X$ is a differentiable function
- X is a complex manifold and $f : X \to X$ is a complex analytic (also called *holomorphic*) function
- X carries a Poisson structure and $f : X \to X$ is a canonical (also called *symplectic*) transformation

Each of these cases has evolved into a separate subdiscipline of the theory of dynamical systems.

In another direction, if f is injective (that is, there is only one solution x to the equation $f(x) = y$ for a given y) we can define its inverse. This allows us to extend the definition of an orbit backwards in time:

$$f(x_{-1}) = x_0, \quad x_{-1} = f_{-1}(x_0)$$

and so on.

16.2. Doubling modulo one

Consider the map of the unit interval $[0, 1]$ to itself

$$f(x) = 2x \bmod 1$$

That is, we double the number and keep its fractional part. Clearly, it has a fixed point at $x = 0$.

A simple way to understand this map is to expand every number in base two. We will get a sequence

$$x = 0.a_1 a_2 a_3 \cdots$$

where a_k are either 0 or 1. Doubling x is the same as shifting this sequence by one step:

$$x = a_1.a_2 a_3 \cdots$$

Taking modulo one amounts to ignoring the piece to the left of the binary point:

$$f(0.a_1 a_2 a_3 \cdots) = 0.a_2 a_3 a_3 \cdots$$

This map is not injective: two values of x are mapped to the same value $f(x)$ since the information in the first digit is lost. A fixed point occurs when all the entries are equal: either 0 or 1 repeated. But both of these represent 0 modulo one. (Recall that $0.11111.. = 1.0 = 0$ mod 1.) So we have just the one fixed point.

An orbit of period two is a sequence

$$0.a_1 a_2 a_1 a_2 a_1 a_2 \cdots$$

We need $a_1 \neq a_2$ so that this is not a fixed point.

Thus we get $0.01010101.. = \frac{1}{3}$ and $0.101010 \cdots = \frac{2}{3}$ which are mapped into each other to get an orbit of period two. Alternatively, they are fixed points of the iterate

$$f_2(x) = f(f(x)) = 2^2 x \bmod 1$$

More generally we see that

$$f_n(x) = 2^n x \bmod 1$$

which has fixed points at

$$x = \frac{k}{2^n - 1}, \quad k = 1, 2, 3, \cdots$$

There are a countably infinite number of such points lying on periodic orbits.

Every rational number has a binary expansion that terminates with a repeating sequence. Thus they lie on orbits that are attracted to some periodic orbit. Since the rational numbers are countable, there are a countably infinite number of such orbits with a predictable behavior. But the irrational numbers in the interval $[0, 1]$ which are a much bigger (i.e., uncountably infinite) set, have no pattern at all in their binary expansion: they have a chaotic but deterministic behavior. This is one of the simplest examples of a chaotic dynamical system.

16.3. Stability of fixed points

If the function is differentiable, we can ask whether a fixed point is stable or not, i.e., whether a small change in initial condition will die out with iterations or not.

Consider again $f : [0, 1] \to [0, 1]$. Suppose x^* is a fixed point

$$f(x^*) = x^*$$

Under a small change from the fixed point

$$x = x^* + \epsilon$$

Then

$$f(x^* + \epsilon) = x^* + \epsilon f'(x^*) + O(\epsilon^2)$$

$$f_2(x^* + \epsilon) = x^* + \epsilon f_2'(x^*) + O(\epsilon^2)$$

But

$$\frac{d}{dx} f(f(x)) = f'(f(x))f'(x)$$

by the chain rule. At a fixed point

$$f_2'(x^*) = f'(x^*)f'(x^*) = [f'(x^*)]^2$$

More generally

$$f_n'(x^*) = [f'(x^*)]^n$$

Thus the distance that a point near x^* moves after n iterations is

$$|f_n(x^* + \epsilon) - x^*| = [f'(x^*)]^n \epsilon + O(\epsilon^2)$$

This will decrease to zero if

$$|f'(x^*)| < 1$$

This is the condition for a *stable fixed point*. On the other hand, if

$$|f'(x^*)| > 1$$

we have an *unstable fixed point*. The case

$$|f'(x^*)| = 1$$

is *marginal*: we have to go to higher orders to determine the behavior near x^*.

Example 16.1: Suppose $f(x) = x^2$ is a map of the closed interval $[0, 1]$. Then 0 is a fixed point. It is stable, as $f'(0) = 0$. But the fixed point at $x = 1$ is unstable as $f'(1) = 2 > 1$. The orbit of 0.99 is driven to zero:

0.99, 0.9801, 0.960596, 0.922745, 0.851458, 0.72498, 0.525596, 0.276252, 0.076315, 0.00582398, 0.0000339187, $1.15048*10^{-9}$, $1.3236*10^{-18}$, $\cdots$

Example 16.2: The map $f(x) = \frac{1}{4}x(1 - x)$ has a stable fixed point at the origin. Where is its other fixed point? Is it stable?

Example 16.3: But for the case $f(x) = 2x(1 - x)$ the origin is an unstable fixed point. It has another fixed point at 0.5 which is stable. If we start near $x = 0$ we will be driven away from it towards $x = 0.5$. For example, the orbit of $x_0 = 10^{-6}$ is

0.00100, 0.001998, 0.00398802, 0.00794422, 0.0157622, 0.0310276, 0.0601297, 0.113028, 0.200506, 0.320606, 0.435636, 0.491715, 0.499863, 0.5, 0.5, 0.5, 0.5, 0.5, 0.5, 0.5, 0.5, 0.5, 0.5, 0.5, 0.5, 0.5, 0.5, 0.5, 0.5, 0.5, $\cdots$

A convenient way of displaying the orbit is a *cobweb* diagram (e.g., Fig. 16.1). Draw a graph of the function $f(x)$ as well as the identity function, which is the diagonal line $y = x$. Start at the point (x_0, x_1) on the graph of the function, where $x_1 = f(x_0)$. Draw a horizontal line to the point on the diagonal (x_1, x_1). Draw a vertical line to the next point on the graph of the function (x_1, x_2), where $x_2 = f(x_1)$. Repeat by drawing the horizontal line to (x_2, x_2) on the diagonal, a vertical line to (x_2, x_3) and so on.

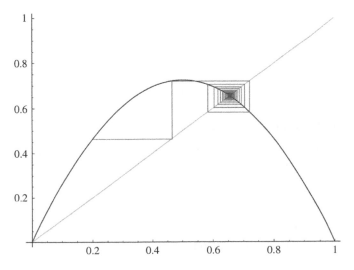

Fig. 16.1 The cobweb diagram of $f(x) = 2.9x(1 - x)$ with initial point $x_0 = 0.2$.

16.4. Unimodal maps

Differentiable maps of the unit interval $f : [0, 1] \to [0, 1]$ that satisfy the boundary conditions $f(0) = f(1) = 0$ must have a maximum between 0 and 1. If there is exactly one such maximum they are said to be unimodal. Examples are

$$f(x) = 3x(1 - x), \quad b \sin \pi x, \text{ for } b < 1$$

etc. They have an interesting dynamics. It turns out that the special case

$$f(x) = \mu x(1 - x)$$

for various values of the parameter μ represents this class of maps very well. Many of the properties are *universal*: independent of the particular unimodal map chosen. So a simple example will suffice.

An interpretation of this system is that this map represents the time evolution of the population of some family of rabbits. Let x be the number of rabbits as a fraction of the maximum number that can be supported by its carrot patch. If x is too close to the maximum value, the number of rabbits in the next generation will be small: many will die of starvation. But if x is small, the next generation will have a number proportional to x, the proportionality constant μ being a measure of the fertility of rabbits.[1] More seriously, it is possible to construct a non-linear electrical circuit (an analog simulator) which implements this dynamics.

We will choose $0 < \mu < 4$ so that the maximum value remains less than one: otherwise the value $f(x)$ might be outside the interval.

A stable fixed point would represent a self-sustaining population. There are at most two fixed points:

$$f(x) = x \implies x = 0, \ 1 - \frac{1}{\mu}$$

Note that

$$f'(0) = \mu, \quad f'\left(1 - \frac{1}{\mu}\right) = 2 - \mu$$

16.4.1 One stable fixed point: $1 > \mu > 0$

When $\mu < 1$, the second fixed point is outside the interval, so it would not be allowed. In this case, the fixed point at the origin is stable:

$$f'(0) = \mu < 1$$

Every point on the interval is attracted to the origin. The fertility of our rabbits is not high enough to sustain a stable population.

---

[1] This is crude. But then, it is only an attempt at quantitative biology and not physics.

16.4.2   One stable and one unstable fixed point: $1 < \mu < 3$

When $1 < \mu < 3$, the fixed point at the origin is unstable while that at $1 - \frac{1}{\mu}$ is stable. A small starting population will grow and reach this stable value after some oscillations. For example, when $\mu = 2.9$ this stable value is 0.655172. A population that is close to 1 will get mapped to a small value at the next generation and will end up at this fixed point again.

16.4.3   Stable orbit of period two : $3 < \mu < 1 + \sqrt{6}$

Interesting things start to happen as we increase μ beyond 3. Both fixed points are now unstable, so it is not possible for the orbits to end in a steady state near them. A periodic orbit of period two occurs when μ is slightly larger than 3 (see Fig. 16.2). That is, there is a solution to

$$f(f(x)) = x$$

within the interval. For example, when $\mu = 3.1$ the solutions are

$$0, \ 0.558014, \ 0.677419, \ 0.764567$$

The first and the third are the unstable fixed points of f that we found earlier. The other two values mapped into each other by f and formed a periodic orbit of period two.

$$f(x_1) = x_2, \quad f(x_2) = x_1$$

$$x_1 \approx 0.558014, \quad x_2 = 0.764567$$

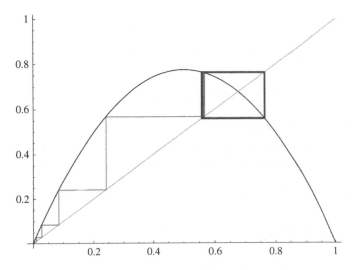

Fig. 16.2 The cobweb diagram of $f(x) = 3.1x(1 - x)$ showing a cycle of period 2.

Is this stable? That amounts to asking whether $|f_2'| < 1$ at these points. If $f(x_1) = x_2$, $f(x_2) = x_1$

$$f_2'(x_1) = f'(f(x_1))f'(x_1) = f'(x_2)f'(x_1)$$

Clearly $f_2'(x_2)$ is equal to the same thing. So we are asking of the product of the derivatives of f along the points on one period is less than one. Numerically, for $\mu = 3.1$ we get $f_2'(x_1) = 0.59$. This means that the orbit of period two is stable. The population of rabbits ends up alternating between these values forever, for most starting values.

We can understand this analytically. To find the fixed points of f_2 we must solve the quartic equation $f_2(x) = x$. We already know two roots of this equation (since 0, $1 - \frac{1}{\mu}$ are fixed points of f hence of f_2). So we can divide by these (that is, simplify $\frac{f_2(x)-x}{x(x-1+\frac{1}{\mu})}$) and reduce the quartic to a quadratic:

$$1 + \mu - x\mu - x\mu^2 + x^2\mu^2 = 0$$

The roots are at

$$x_{1,2} = \frac{1 + \mu \pm \sqrt{-3 - 2\mu + \mu^2}}{2\mu}$$

As long as $\mu > 3$ these roots are real: we get a periodic orbit of period two. We can calculate the derivative at this fixed point as above:

$$f_2'(x_1) = 4 + 2\mu - \mu^2$$

For the period two orbit to be stable we get the condition (put $f_2'(x_1) = -1$ to get the maximum value for stability)

$$\mu < 1 + \sqrt{6} \approx 3.44949$$

16.4.4   Stable orbit of period four : $3.44949\ldots < \mu < 3.54409\ldots$

Thus, as μ increases further this orbit will become unstable as well: a period four orbit develops (see Fig. 16.3). Numerically, we can show that it is stable till $\mu \approx 3.54409\ldots$

16.4.5 $3.54409\ldots < \mu < \mu_\infty$

After that, there is a stable period eight orbit until $\mu \approx 3.5644\ldots$ and so on. Let μ_n be the value at which a period 2^n orbit first appears. They form a convergent sequence:

$$3,\ 3.44949,\ 3.54409,\ 3.5644,\ 3.568759,\ \rightarrow \mu_\infty \approx 3.568856$$

The rate of convergence is geometric. Using much more sophisticated methods, Feigenbaum (1978) obtained

$$\lim_{n\to\infty} \frac{\mu_n - \mu_{n-1}}{\mu_{n+1} - \mu_n} \equiv \delta = 4.669\ldots$$

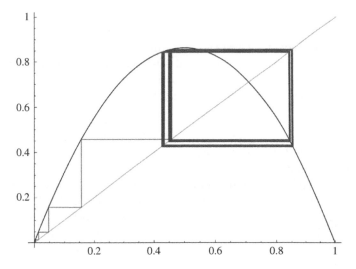

Fig. 16.3 The cobweb diagram of $f(x) = 3.4495x(1 - x)$ showing a cycle of period 4.

A study of another unimodal map such as $f(x) = \mu \sin \pi x$ (restricted to the unit interval) will lead to a similar, but numerically different, sequence. While studying this, Feigenbaum made a surprising discovery: although the values at which the period doubles depend on the choice of f, *the rate of convergence is universal*: it is always the same mysterious number 4.669... This reminded him of similar universality in the theory of critical phenomena (such as when a gas turns into a liquid) which had been explained by Wilson using an esoteric theoretical tool called *renormalization*. He then developed a version of renormalization to explain this universality in chaos, leading to the first quantitative theory of chaos. In particular, the universal ratio above turns out to be the eigenvalue of a linear operator.

16.5. Invariant measures

An important question raised by Boltzmann in the nineteenth century is whether Hamiltonian systems that are far from being integrable (e.g., the dynamics of a large number of molecules) are ergodic. Is the fraction of the time that an orbit spends in a domain of the phase space[2] proportional to its volume? Is this true for almost every orbit? If these hypotheses hold, we would be able to derive statistical mechanics from Hamiltonian mechanics. We are far from a general resolution of these questions. But much progress has been made in the past century and a half. Many special case have been shown to be ergodic. In the other direction, the Kolmogorov–Arnold–Moser theorem shows that nearly integrable systems are not.

---

[2]More precisely, the submanifold of states of a given energy.

To talk of volumes in phase space, we need a measure of integration that is invariant under time evolution. We saw that for a Hamiltonian system with a non-degenerate Poisson bracket, there is an invariant measure (volume) in phase space. In canonical co-ordinates, this is just $\prod_i dp_i dq_i$. For general dynamical systems, there may not always be such a measure:[3] the system could be dissipative, so that all orbits concentrate on some small subset (even a point). But in many cases we can find such an invariant measure.

Example 16.4: For the doubling map $f(x) = 2x$ mod 1, the usual Lebesgue measure dx on the unit interval is invariant. The reason is that f is a two-to-one map, with Jacobian 2. Put differently, each interval arises as the image of two others, each with half the length of the original.

It turns out that the doubling map is indeed ergodic on the unit interval with respect to this measure. Suppose A is an invariant set of f. That is, if $x \in A$ then $f(x) \in A$ as well. An example would be a periodic orbit. This example has zero measure: A is contained in unions of intervals that can be made as small as you want. There are also invariant sets of positive measure. For example, the orbit of an interval $[0, \epsilon]$.

$$A = \bigcup_{n=-\infty}^{\infty} f^n([0, \epsilon])$$

It can be shown that any invariant set of f with positive measure is of full measure. That is, a set invariant under the dynamics is either of volume zero or its volume is equal to that of the whole interval. This is a form of ergodicity.

This is just the beginning of a vast area of ergodic theory.

Problem 16.1: Show that for the fully chaotic logistic map $f(x) - 4x(1 - x)$, the measure

$$d\mu(x) = \frac{dx}{\pi\sqrt{x(1 - x)}}$$

is invariant. (Remember that f is a two-to-one map.)

Problem 16.2: Plot the periodic cycles of the map $f(x) = \mu \sin(\pi x)$ letting μ vary in the range $0 < \mu < 1$ in small steps (about 10^{-2}). For each value of μ, start with some value of x (e.g., near 0.5) and make a list of a few hundred points in its orbit, after dropping the first one hundred or so. This should produce a cycle (see Fig. 16.4). Plot the list of values (μ, x) you get this way.

Problem 16.3: Study the dynamics of the map $f(x) = \mu x(1 - x)$ for the parameter range $0 < \mu < 4$ numerically. Find the first four values of μ at which the period doubles.

Problem 16.4: *Project*** We can apply the idea of iteration to the space of unimodal functions itself, by repeating the above construction. The fixed point

[3]There may not be an invariant measure that is absolutely continuous w.r.t. the Lebesgue measure. It might be concentrated at some points, for example.

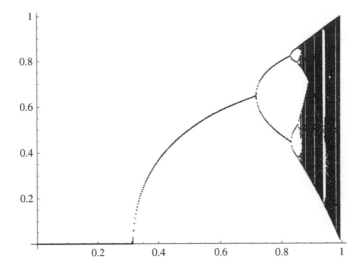

Fig. 16.4 The cycles of the map $f(x) = \mu \sin(\pi x)$ for the range $0 < \mu < 1$.

will be a universal unimodal function which has an orbit of infinite period. The constant δ in Section 16.4.5 can be understood as the rate of convergence to this fixed point. Read Feigenbaum's article (1978) and verify his results.

17
Dynamics on the complex plane

There are two natural groups of transformations of the plane. If the areas are preserved we get canonical (also called symplectic) transformations. If the angles of intersection of any pair of lines are preserved, we get conformal transformations. These are simply complex analytic functions. If both areas and angles are preserved, the transformation is an isometry of the Euclidean metric: a combination of a rotation and a translation. We have already studied iterations of canonical transformations. In this chapter, we study the iterations of a complex analytic function of one variable. For a deeper study, see the books by Milnor (1999), and McMullen (1994).

17.1.   The Riemann sphere

The simplest analytic functions are polynomials. They are analytic everywhere on the complex plane. At infinity, a polynomial must have infinite value, except in the trivial case where it is a constant. It is convenient to extend the complex plane by including the point at infinity, $\hat{\mathbb{C}} = \mathbb{C} \cup \{\infty\}$. A rational function is of the form $R(z) = \frac{P(z)}{Q(z)}$ where $P(z)$ and $Q(z)$ are polynomials (without common factors, because they could be cancelled out). Rational functions can be thought of as analytic functions mapping $\hat{\mathbb{C}}$ to itself: whenever the denominator vanishes, its value is the new point at infinity we added. For example, the function $\frac{1}{z-1}$ can now be thought of as an analytic function that maps ∞ to zero and 1 to ∞.

An important geometric idea is that $\hat{\mathbb{C}}$ can be identified with a sphere. More precisely, there is a coordinate system (the stereographic system) on $\mathbb{S}^2$ that associates to every point p on it a complex number; place the sphere on the plane so that its south pole touches the origin. Draw a straight line from the north pole to $p \in \mathbb{S}^2$; continue it until it reaches the plane at some point $z(p)$. This is the co-ordinate of p. Clearly the co-ordinate of the south pole is zero. The equator is mapped to the unit circle. As we get close to the north pole, the co-ordinate gets larger: the north pole itself is mapped to the point at infinity.

A moment's thought will show that this map is invertible: to every point on $\hat{\mathbb{C}}$ there is exactly one point on $\mathbb{S}^2$. So the sphere is nothing but the complex plane with the point at infinity added. This point of view on the sphere is named for Riemann, the founder of complex geometry as well as Riemannian geometry.

Thus, rational functions are complex analytic maps of the sphere to itself. The study of the dynamics of such maps is an interesting subject. Even simple functions such as $f(z) = z^2 + c$ (for constant c) lead to chaotic behavior. But we begin with a simpler case that is not chaotic.

17.2. Mobius transformations

A rotation takes the sphere to itself in an invertible way: each point has a unique inverse image. It preserves the distances between points. In complex geometry, there is a larger class of analytic maps that take the sphere to itself invertibly. The rotations are a subset of these. To determine these we ask for the set of rational functions for which the equation

$$R(z) = w$$

has exactly one solution z for each w. This is the same as solving the equation

$$P(z) - wQ(z) = 0, \quad R(z) = \frac{P(z)}{Q(z)}$$

The number of solutions is the degree of the polynomial $P(z) - wQ(z)$ in the variable z. This is the larger of the degrees of $P(z)$ or $Q(z)$: which is called the degree of the rational function $R(z)$. Thus, for example, $\frac{1}{z^2+3}$ has degree two. So we see that invertible maps of the sphere correspond to rational functions of degree one; i.e., $P(z)$ and $Q(z)$ are both linear functions

$$M(z) = \frac{az + b}{cz + d}$$

But, the numerator and denominator should not be just multiples of each other: then $M(z)$ is a constant. To avoid this, we impose

$$ad - bc \neq 0$$

Exercise 17.1: Check that $ad - bc = 0$, if and only if $R(z)$ is a constant.
In fact, by dividing though by a constant, we can even choose

$$ad - bc = 1$$

We are interested in iterating maps, so let us ask for the composition of two such maps $M_3(z) = M_1(M_2(z))$. We will often denote this composition of maps by $\circ$:

$$M_3 = M_2 \circ M_1$$

It must also be a ratio of linear functions: invertible functions compose to give invertible functions. Calculate away to check that

$$M_i(z) = \frac{a_i z + b_i}{c_i z + d_i}, \quad i = 1, 2, 3$$

have coefficients related by matrix multiplication:

$$\begin{pmatrix} a_3 & b_3 \\ c_3 & d_3 \end{pmatrix} = \begin{pmatrix} a_1 & b_1 \\ c_1 & d_1 \end{pmatrix} \begin{pmatrix} a_2 & b_2 \\ c_2 & d_2 \end{pmatrix}$$

Thus Mobius transformations correspond[1] to the group of 2×2 matrices of determinant one, also called $SL(2, C)$. Rotations correspond to the subgroup of unitary matrices.

It turns out that the general Mobius transformation is closely related to a Lorentz transformation in space-time. The set of light rays (null vectors) passing through the origin in Minkowski space is a sphere. A Lorentz transformation will map one light ray to another: this is just a Mobius transformation of the sphere.

If we ask how the ratio of two components of a vector transforms under a linear transformation, we get a Mobius transformation

$$\begin{pmatrix} \psi_1 \\ \psi_2 \end{pmatrix} \mapsto \begin{pmatrix} a & b \\ c & d \end{pmatrix} \begin{pmatrix} \psi_1 \\ \psi_2 \end{pmatrix}$$

$$\frac{\psi_1}{\psi_2} \equiv z \mapsto \frac{az + b}{cz + d}.$$

This gives another, algebraic interpretation.

17.3. Dynamics of a Mobius transformation

We can now ask for the effect of iterating a Mobius transformation $M(z) = \frac{az+b}{cz+d}$. Given an initial point z_0, we get an infinite sequence

$$z_n = M(z_{n-1}), \quad n = 0, 1, 2, \cdots$$

Because M is invertible, we can also go backwards: define z_{-1} as the solution of the equation $M(z_{-1}) = z_0$, and similarly $z_{-2} = M^{-1}(z_{-1})$ and so on. Thus the orbit of a point is a sequence

$$z_n = M(z_{n-1}), \quad n = \cdots, -2, -1, 0, 1, 2, \cdots$$

which is infinite in both directions.

Let us find fixed points of $M(z)$:

$$M(z) = z \implies az + b = z(cz + d)$$

There are two possible roots for this quadratic equation

$$z_\pm = \frac{a - d \pm \sqrt{(a - d)^2 + 4bc}}{2c}.$$

---

[1]The alert reader will notice that this is a pretty little lie. Mobius transformations actually correspond to $PSL(2, C) = SL(2.C)/\{1, -1\}$, as reversing the signs of all the components of the matrix does not affect $M(z)$. What is the subgroup of rotations?

The quantity $\Delta = (a-d)^2 + 4bc$ is called the *discriminant*. If $\Delta \neq 0$, the Mobius transformation $M(z)$ has two distinct fixed points. Since $ad - bc = 1$, we also have $\Delta = (a+d)^2 - 4$.

Next, we ask if each fixed point is stable. That is, whether $|f'(z_\pm)| < 1$. Using again the fact that $ad - bc = 1$ we find that $f'(z) = \frac{1}{(cz+d)^2}$. The solution above gives

$$f'(z_\pm) = \frac{a+d \mp \sqrt{\Delta}}{2}$$

The derivatives are reciprocals of each other:

$$f'(z_+)f'(z_-) = 1$$

So, if one fixed point is stable, the other unstable. It is also possible for f' to be reciprocal numbers of magnitude one at the fixed points. Then the fixed points are neutrally stable. Finally, if the fixed points coincide, f' must be 1.

Collecting the above facts, we see that there are three classes of Mobius transformations:

1. *Hyperbolic* (also called loxodromic) Two distinct fixed points, where one is stable and the other unstable. This is the generic case. That is, six independent real numbers are needed to determine a hyperbolic element: the three complex numbers determining the positions of the fixed points and the derivative at one of them.
2. *Elliptic* Two distinct fixed points again, and each is marginally stable. Since the magnitude of the fixed point must be one, there is a five-real-parameter family of them.
3. *Parabolic* There is only one fixed point, at which the derivative is one. There is a four-parameter family of them: the co-ordinates of the fixed point and the value of the complex number μ which determines the behavior at a fixed point (see below).

Example 17.1: Here are some examples:

- *Hyperbolic*

$$\begin{pmatrix} 2 & 0 \\ 0 & \frac{1}{2} \end{pmatrix}, \begin{pmatrix} 2 & e^{i\theta} \\ e^{-i\theta} & 1 \end{pmatrix}, \quad 0 < \theta < 2\pi$$

- *Elliptic*

$$\begin{pmatrix} e^{i\theta} & b \\ 0 & e^{-i\theta} \end{pmatrix}, \quad 0 < \theta < 2\pi$$

- *Parabolic*

$$\begin{pmatrix} 1 & 0 \\ 4 & 1 \end{pmatrix}$$

17.3.1   The standard form of a Mobius transformation

We say that two Mobius transformations M_1, M_2 are equivalent if there is another Mobius transformation S such that

$$M_2 = S \circ M_1 \circ S^{-1}$$

S will map the fixed points of M_1 to those of M_2. If M is not parabolic, we can find an S that moves one of the fixed points to the origin and the other to infinity; just choose $S = \frac{1}{\sqrt{z_- - z_+}} \begin{pmatrix} z_- & z_+ \\ 1 & 1 \end{pmatrix}$. If M_1 is parabolic, we can choose S so that the fixed point of M_2 is at infinity. This way we can bring a Mobius transformation M to a standard form in each case:

1. *Hyperbolic* $S \circ M \circ S^{-1} = \begin{pmatrix} \lambda & 0 \\ 0 & \lambda^{-1} \end{pmatrix}$, $|\lambda| < 1$, $z \mapsto \lambda^2 z$

2. *Elliptic* $S \circ M \circ S^{-1} = \begin{pmatrix} e^{i\frac{\theta}{2}} & 0 \\ 0 & e^{-i\frac{\theta}{2}}, \end{pmatrix}$, $0 < \theta < 2\pi$. $z \mapsto e^{i\theta} z$

3. *Parabolic* $S \circ M \circ S^{-1} = \begin{pmatrix} 1 & 0 \\ \mu & 1 \end{pmatrix}$, $z \mapsto \frac{z}{\mu z + 1}$

This is useful because it is easy to determine the orbits of these standard forms. We can then transform back to determine the orbits of the original M.

The orbits of $z \mapsto \lambda^2 z$ are spirals converging to the origin. When $\lambda < 1$, the origin is stable and infinity is unstable. The transformation $z \mapsto e^{i\theta} z$ is a rotation around the origin. The orbits lie along a circle. The parabolic transformation $z \mapsto \frac{z}{\mu z + 1}$ has orbits converging to the origin.

Once we know the orbits of the standard form, we can determine those of the general case by applying S. Or we can simply plot some examples numerically.

17.4.   The map $z \mapsto z^2$

Let us look at the simplest map of degree 2

$$z \mapsto z^2$$

Clearly, any point in the interior of the unit circle will get mapped to the origin after many iterations. Any point outside the unit circle will go off to infinity. On the unit circle itself, we get the doubling map we studied in the last chapter:

$$z = e^{2\pi i x}, \quad x \mapsto 2x \bmod 1$$

We saw that the latter is chaotic: an irrational value of x has a completely unpredictable orbit. But these are most of the points on the unit circle. Even the periodic orbits (rational numbers with a denominator that is a power of two) have something wild about them: they are all unstable. The unit circle is the union of such unstable periodic points. This is a signature of chaotic behavior.

Thus the complex plane can be divided into two types of points:. Those not on the unit circle, which get mapped predictably, and those on the unit circle whose dynamics is chaotic. We will see that points on the plane can always be classified into two such complementary subsets. The chaotic points belong to the *Julia set* and the rest belong to the *Fatou set*. But, in general, the Julia set is not necessarily something simple like the unit circle. Often, it is a quite intricate fractal.

For example, the Julia set of the map $z \mapsto z^2 + i$ is a "dendrite": a set with many branches each of which are branched again. Why does this happen? The key is that at its fixed point (which is unstable), the derivative is not a real number. That is, near the fixed point, the map is both a scaling (stretching) and a rotation. The orbits are spirals. Such a set cannot lie in a smooth submanifold of the plane (unless it consists of the whole plane). So, whenever we have an unstable fixed point with a derivative that is not real (and we know that the domain of repulsion is not the whole plane) the Julia set will be some kind of fractal. This is very common.

How do we come up with a definition of when an orbit is chaotic and when it is not? That is, which points belong to the Julia set and which to the Fatou set? A more precise definition is needed to make further progress.

We will need a notion of convergence of maps to make sense of this idea; and also a notion of distance on the Riemann sphere that is invariant under rotations.

17.5.   Metric on the sphere

In spherical polar co-ordinates, the distance between neighboring points on the sphere is

$$ds^2 = d\theta^2 + \sin^2\theta d\phi^2$$

The stereographic co-ordinates are related to this by

$$z = \cot\frac{\theta}{2}e^{i\phi}$$

Rewritten in these co-ordinates the metric is

$$ds^2 = \frac{4|dz|^2}{(1+z\bar{z})^2}$$

The distance s between two points z_1, z_2 can be calculated from this by integration along a great circle. For our purposes, an equally good notion of distance in the length d of the chord that connects the two points. Some geometry will give you the relation between the two notions:

$$d = 2\sin\frac{s}{2}$$

$$d(z_1, z_2) = \frac{2|z_1 - z_2|}{\sqrt{(1 + |z_1|^2)(1 + |z_2|^2)}}$$

The advantage of these notions of distance over the more familiar $|z_1 - z_2|$ is that they are invariant under rotations. For example, the point at infinity is at a finite distance:

$$d(z, \infty) = \frac{2}{\sqrt{(1 + |z|^2)}}$$

We only care about the topology defined by the distance, which is the same for d and s: that is, when one is small the other is also. So any sequence that converges in one will converge in the other. It is a matter of convenience which one we use in a computation.

17.6. Metric spaces

Metric is another word for distance. A *metric* d on a set X is a function $d : X \times X \to \mathbb{R}$ such that

1. $d(x, x) = 0$
2. $d(x, x') > 0$ if $x \neq x'$ Positive
3. $d(x, x') = d(x', x)$ Symmetry
4. $d(x, x') \leq d(x, x'') + d(x'', x')$ The triangle inequality

The most obvious example is the case of Euclidean space $X = \mathbb{R}^n$ and

$$d(x, x') = |x - y| = \sqrt{\sum_{i=1}^{n} (x^i - x^{i\prime})^2}$$

The chordal distance (the length of the chord connecting two points) on the sphere is another example of a metric.

Exercise 17.2: Prove the triangle inequality for the chordal distance.

The metric is a useful concept because it allows us to speak of convergence. We say that $x_n \in X$ converges to x if for every ϵ there is an N such that $d(x_n, x) < \epsilon$ for $n \geq N$. It also allows us to speak of a continuous function. A function $f : X \to Y$ from one metric space to another is continuous if for every $\epsilon > 0$ there is a δ such that

$$d_Y(f(x), f(x')) < \delta, \text{ for } d_X(x, x') < \epsilon$$

That is, we can make the values of a function at two points as small as we want by bring the points close enough.

17.6.1   Compactness

Not every sequence converges. The points might move farther and farther away from the initial one as n grows. Or they might wander, approaching one point then another and back to the first and so on. In the latter case, there would be a subsequence with a limit. A compact set is one in which every sequence has a convergent subsequence.

Continuous functions on compact sets have many of the properties that a function on a finite set would have. For example, a real-valued function on a compact set is bounded. This is why it is so important: compactness allows us to treat some very large, infinite, sets as if they are finite. It makes sense to speak of the supremum (roughly speaking the maximum value; more precisely the least upper bound) or the infimum (least greatest lower bound) of a function on a compact set. These notions do not make sense if the domain is non-compact.

> **Example 17.2:** The reciprocal is a continuous function on the open interval (0, 1). It does not have a supremum: it grows without bound near 0. Thus, the open interval is not compact. On the other hand, the closed interval $[0,1] = \{x|0 \le x \le 1\}$ is compact. Any continuous function on it is bounded. The reciprocal is not a continuous function on the closed interval.

17.6.2   Distance between functions

Often we will need to ask whether a sequence of functions has a limit. (For example, does a power series converge in some domain? This is the same as asking if the sequence of partial sums converge to some function.) It is useful in this case to have a notion of distance between functions. If the domain K is compact, we can take the largest distance between the values of the functions and declare that to be the distance between functions themselves

$$d(f,g) = \sup_{x \in K} d(f(x), g(x))$$

This turns the set of functions into a metric space as well, allowing us to speak of convergence of a sequence of functions. A *normal family* is a set of functions such that every infinite sequence in it has a convergent subsequence. In other words, it is a compact subset of the metric space of functions.

17.7.   Julia and Fatou sets

When do the iterates $f, f^2, f^3, \cdots$ of an analytic function f tend to a limit ϕ that is also an analytic function? This question is important because if such a limiting function exists, the dynamics would be easy to understand: it would not be chaotic.

This is where we will use the idea of convergence of sequences of functions from the previous section. So we define the *Fatou set* of f to be the largest domain on which the iterates f_n have a subsequence that converges to an analytic function ϕ. That is the largest domain on which the iterates form a normal family. The complement of the Fatou set is the *Julia set*. On it, the dynamics are chaotic. Here are some theorems about the Julia set that help to explain what it is. Look at the books by Milnor (1999) and McMullen (1994) for proofs and a deeper study.

Theorem 17.1: *Let J be the Julia set of some analytic function $f : S^2 \to S^2$. Then, $z \in J$, if and only if its image $f(z) \in J$ as well.*

Thus, the Julia set is invariant under the dynamics; the Julia set of f and those of its iterates f^n are the same.

Theorem 17.2: *Every stable periodic orbit is in the Fatou set. However, every unstable periodic orbit is in the Julia set.*

Recall that if z is in a periodic orbit of period k, we have $f^k(z) = z$. It is stable if $|f^{k\prime}(z)| < 1$ and unstable if $|f^{k\prime}(z)| > 1$. There is a neighborhood of every periodic orbit in which all points will get closer and closer to the periodic orbit as we iterate f. This is called the basin of attraction of the orbit.

Theorem 17.3: *The basin of attraction of every stable periodic point is in the Fatou set. Roughly speaking, the Fatou set consists of all the stable periodic orbits and their basins of attraction. On the other hand...*

Theorem 17.4: *The Julia set is the closure of the set of unstable periodic orbits. This means that every point in the Julia set can be approximated as close as we want by some unstable periodic point.*

It is useful to have a characterization of the Julia set that allows us to compute it.

Theorem 17.5: *If z_0 is in the Julia set J of some analytic function f, then the set of pre-images of z_0 is everywhere dense in J.*

That is, the Julia set is well-approximated by the set

$$\{z \in S^2 : f^n(z) = z_0,\ n \geq 1\}$$

where f^n is the nth iterate of f. So to compute a Julia set, we must first find one point z_0 on it. (For example, pick an unstable fixed point). Then we solve the equation $f(z) = z_0$ to find the first set of pre-images. If $f(z)$ is a rational function, this amounts to solving some polynomial equation, which has some finite number of solutions. Then we find all the pre-images of these points and so on. The number of points grows exponentially. If we plot them, we get an increasingly accurate picture of the Julia set. The point in doing this "backwards evolution" is that it is stable. Precisely because f is unstable near z_0, its inverse is stable: we can reliably calculate the solutions to $f(z) = z_0$ since the errors in z_0 will be damped out in the solutions.

In practice, it might make sense to pick just one pre-image at random at each step. This way, we can search in depth and not be overwhelmed by the exponential growth of pre-images. Although the picture of the Julia set you get this way is not as pretty as some others you will see, it is more representative of the dynamics. Areas that are visited often will be darker (see Fig. 17.2). For more on these matters see Milnor (1999).

Problem 17.3: What is the relation between the fixed points of a Mobius transformation and the eigenvectors of the corresponding matrix? Same for the eigenvalues and the derivatives of the Mobius transformation at a fixed point.

Compare the classification of Mobius transformations to Jordan canonical forms of matrices. When can a matrix *not* be diagonalized?

Problem 17.4: Find a Mobius transformation that has fixed points at given points z_u, z_s and has derivative at z_s equal to λ. Plot some orbits of the hyperbolic Mobius transformation with $z_u = 2.0 + 2.1i$, $z_s = -8 - 10.1\,i$, $\lambda = 0.9896 - 0.0310997 * i$. Plot orbits of the parabolic Mobius transformation (see Fig. 17.3) $z \mapsto \frac{z}{\mu z + 1}$ for some values of μ.

Problem 17.5: Write a program that will plot the Julia set of the map $P(z) = z^2 + c$ for different choices of c, such as those plotted in Figure 17.1.

Problem 17.6: *The Mandelblot* The Mandelbrot set is the set of all values of the parameter c for which the orbit of 0 under the above map remains bounded

$$z_n(c) = z_{n-1}^2(c) + c, \quad z_0(c) = 0$$
$$M = \{c | \exists s > 0, \forall n, |z_n(c)| < s\}$$

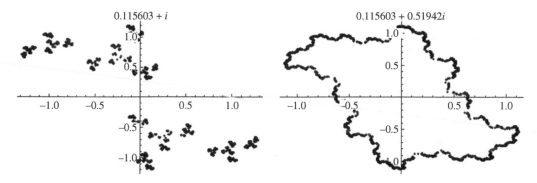

Fig. 17.1 Julia sets for $z^2 + c$ for various values of c.

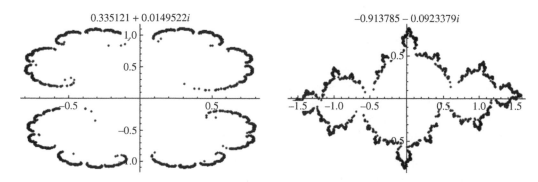

Fig. 17.2 More Julia sets.

Fig. 17.3 Orbits of a hyperbolic Mobius transformation.

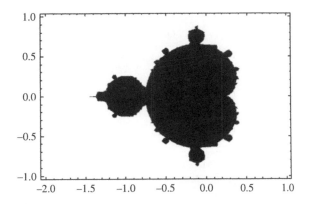

Fig. 17.4 The Mandelbrot set.

This set has a visually amazing fractal structure. Write a program that plots the Mandelbrot set (Fig. 17.4). You can also color code it by the number of iterations it takes for the orbit to leave the range of the plot.

18
KAM theory

The Kolmogorov–Arnold–Moser (KAM) theorem of mechanics places limits on when chaos can set in. It gives conditions under which an integrable system can be perturbed and remain predictable. This is a deep theory with many mathematical difficulties. So we will work up to it by solving a couple of much easier problems in this chapter. We will use Newton's ancient iteration method to solve the problem of diagonalizing matrices, and to bring diffeomorphisms of the circle to normal form. The first doesn't really have much to do with chaos; it is only included as a warm-up exercise. The second is a toy model for mechanics. More detailed exposition can be found in Wayne (2008), Celletti and Chierchia (2007) and Arnold (1961 and 1989).

The essential difficulty to overcome is that of small denominators in perturbation theory. If you are close to a resonance, a small perturbation can have a large effect, in the linear approximation. More precisely, the measure of the size of a perturbation must take into account the effect of possible resonances. Once the correct measure of the size of a perturbation is found, a Newton's iteration solves the problem.

18.1. Newton's iteration

One of the earliest uses of iterations is to solve non-linear equations. Suppose we want to solve an equation $f(\alpha) = 0$ in one variable (real or complex). Given a guess x for the root, Newton's method is to solve the first order approximation

$$0 = f(\alpha) \approx f(x) + f'(x)(\alpha - x)$$

to get an improved guess:

$$\alpha \approx x - [f'(x)]^{-1} f(x)$$

We then repeat the whole process. In other words, we iterate the function

$$R(x) = x - [f'(x)]^{-1} f(x)$$

The sequence $x_{k+1} = R(x_k)$ converges if the initial guess x_0 is close enough to a root *and* $f'(x_k)$ does not ever become small.

The method even applies to the case of several variables, except that x_k are now vectors and $f'(x)$ is a matrix. Again, there are problems if the matrix has small eigenvalues; that is, if its inverse becomes large. (The problem of small denominators.) Indeed, this method even generalizes to infinite-dimensional spaces, provided we have a way of measuring the length of a vector and hence a norm for operators; for example, to Hilbert or Banach spaces.

18.1.1   Proof of convergence

We will show that as long as

$$f'(x) \neq 0 \tag{18.1}$$

along the orbit, the Newton iteration will converge.

A measure of how far away we are from a root is given by

$$\delta(x) = |R(x) - x|$$

where

$$R(x) = x - r(x), \quad r(x) = \frac{f(x)}{f'(x)}$$

Under an iteration, it changes to

$$\delta(R(R(x)) - R(x)) = |r(R(x))| = |r(x - r(x))| \approx |r(x)[1 - r'(x)]| = \delta(x)|1 - r'(x)|$$

But

$$1 - r'(x) = \frac{f(x)f''(x)}{f'^2(x)} = \frac{f''(x)}{f'(x)} r(x)$$

Thus we have

$$\delta(x_{k+1}) \approx \left| \frac{f''(x_k)}{f'(x_k)} \right| \delta^2(x_k)$$

so that the error decreases quadratically as long as the criterion 18.1 is satisfied. That is, at each step, the error is roughly the square of the error in the previous step.

18.2.   Diagonalization of matrices

Suppose we are given a matrix L. When is there an invertible linear transformation (similarity transformation) S that will bring L to diagonal form?

$$S^{-1}LS = \Lambda, \quad \Lambda = \begin{pmatrix} \lambda_1 & 0 & \cdots \\ 0 & \lambda_2 & \cdots \\ \cdot & \cdot & \cdots \end{pmatrix} \tag{18.2}$$

Not every matrix can be diagonalized. It is not even true that small perturbations of diagonal matrices can always be diagonalized.

Example 18.1:

$$L = \begin{pmatrix} 1 & \epsilon \\ 0 & 1 \end{pmatrix}$$

cannot be diagonalized for $\epsilon \neq 0$, no matter how small ϵ is.

Small perturbations of diagonal matrices with *unequal diagonal entries* (non-degenerate eigenvalues) can be diagonalized. One approach to this is to use Newton's iteration.

Define a quantity that measures the size of the off-diagonal part

$$\delta(L) = \sum_{i \neq j} \frac{|L_{ij}|}{|L_{ii} - L_{jj}|}$$

Notice that each off-diagonal element is compared to the difference of the corresponding diagonal entries. Thus, a matrix like that in the above example will never have small $\delta(L)$: we would need the diagonal entries to be unequal for that.

We will now construct a similarity transformation that makes $\delta(L)$ smaller, assuming it started off small. As in Newton's method, this is found by solving the linear approximation to the equation 18.2.

Let

$$S = 1 + s, \quad s_{ij} = 0 \text{ for } i = j$$

To first order in s, $S^{-1} \approx 1 - s$. Treating the off-diagonal entries of L as small as well, 18.2 gives

$$L_{ij} \approx (L_{jj} - L_{ii}) s_{ij}, \quad \text{for } i \neq j$$

When $\delta(L)$ is small, $L_{ii} \neq L_{jj}$ and the solution

$$s_{ij} \approx \frac{L_{ij}}{L_{jj} - L_{ii}}; \quad i \neq j, \quad s_{ii} = 0 \tag{18.3}$$

can be found. Moreover, the size of s is small:

$$|s| \equiv \sum_{i \neq j} |s_{ij}| = \delta(L)$$

Thus, we define

$$[S(L)]_{ii} = 1, \quad [S(L)]_{ij} = \frac{L_{ij}}{L_{jj} - L_{ii}}, \quad \text{for } i \neq j$$

Now we define an iteration as in Newton's method

$$R(L) = S(L)^{-1} L S(L)$$

If $\delta(L)$ is small enough, at each step it will become even smaller. Eventually we will get a diagonal matrix to high precision. We skip a proof and instead ask the reader to work out a few examples.

Example 18.2: Even for a matrix with fairly close diagonal entries and not small of-diagonal entries,

$$L = \begin{pmatrix} 1. & 1.63051 \\ 4.75517 & 1.01 \end{pmatrix}, \quad \delta(L) = 212.855$$

we get a sequence of L's

$$\begin{pmatrix} 0.960004 & 1.6303 \\ 4.75455 & 1.05 \end{pmatrix}, \begin{pmatrix} 0.870106 & -1.6286 \\ -4.74959 & 1.13989 \end{pmatrix}, \begin{pmatrix} 0.602832 & 1.61341 \\ 4.70531 & 1.40717 \end{pmatrix},$$

$$\begin{pmatrix} -0.138342 & -1.48672 \\ -4.33582 & 2.14834 \end{pmatrix}, \begin{pmatrix} -1.40089 & 0.820858 \\ 2.39393 & 3.41089 \end{pmatrix}, \begin{pmatrix} -1.77733 & -0.0642183 \\ -0.187284 & 3.78733 \end{pmatrix},$$

converging to

$$\begin{pmatrix} -1.77949 & 0.0000249329 \\ 0.0000727137 & 3.78949 \end{pmatrix}, \quad \delta(L) = 0.0000175341.$$

in just seven iterations.

The point is not that this is a numerically efficient way of diagonalizing a matrix; it may not be. It is a toy model for the much harder problem of bringing non-linear maps to a standard (also called normal) form.

Remark 18.1: A small modification of the above procedure will ensure that the transformation $S(L)$ is unitary when L is hermitian. We choose s as before (18.3) but put

$$S = \left(1 + \frac{1}{2}s\right)\left(1 - \frac{1}{2}s\right)^{-1}$$

18.3. Normal form of circle maps

Consider a map of the circle to itself with $w' > 0$ and

$$w(\theta + 2\pi) = \theta + 2\pi$$

Such a map is called a diffeomorphism: a smooth invertible map of the circle. The simplest example is a rotation

$$w(\theta) = \theta + \rho$$

some constant ρ.

Example 18.3: If $\frac{\rho}{2\pi} = \frac{m}{n}$ for a pair of co-prime integers, the orbit of the rotation $\theta \mapsto \theta + \rho$ is periodic with period n. If $\frac{\rho}{2\pi}$ is irrational, the orbit is dense. That is, every orbit will come as close to a given point as you want.

$$\forall \; \theta, \theta_0 \text{ and } \epsilon > 0, \exists n \text{ such that } |\theta_0 + n\rho - \theta| < \epsilon$$

Is there a change of co-ordinate $\theta \mapsto S(\theta)$ that reduces a generic diffeomorphism to a rotation?

$$S^{-1} \circ w \circ S(\theta) = \theta + \lambda \tag{18.4}$$

This is also called the *normal form* of the diffeomorphism. This is analogous to asking if a matrix can be brought to diagonal form by a change of basis. The number λ is analogous to the spectrum of the matrix.

Example 18.4: The doubling map $\theta \mapsto 2\theta \bmod 2\pi$ cannot be brought to normal form.

This question is important because such a normalizable diffeomorphism cannot be chaotic: up to a change of variables, it is just a rotation. The orbit can be dense, but it will be as predictable as any rotation. Thus, a theorem that characterizes normalizable maps puts limits on chaos. The surprise is that rotations with rational $\frac{\rho}{2\pi}$ are unstable: there are perturbations of arbitrarily small size which are chaotic. If the rotation is "irrational enough", it has a neighborhood in which all the maps are normalizable and hence non-chaotic. The more irrational the rotation is, the larger this neighborhood. We will soon see how to measure the irrationality of a rotation.

Again, we can gain some insight by studying the linear approximation to the problem. We can split $w(\theta)$ as a rotation plus a periodic function of zero mean:

$$w(\theta) = \theta + \rho + \eta(\theta), \quad \int_0^{2\pi} \eta(\theta)d\theta = 0, \quad \eta(\theta + 2\pi) = \eta(\theta)$$

Clearly,

$$\rho = \frac{1}{2\pi} \int_0^{2\pi} [w(\theta) - \theta]d\theta$$

We can similarly split S to the identity plus a correction

$$S(\theta) = \theta + s(\theta), \quad s(\theta + 2\pi) = s(\theta)$$

Treating s and η as small, the first order approximation to 18.4 is

$$s(\theta + \rho) - s(\theta) = \eta(\theta) \tag{18.5}$$

To see this, write 18.4 as $w(\theta + s(\theta)) = \theta + \lambda + s(\theta + \lambda)$. To first order, $\lambda \approx \rho$ and $w(\theta + s(\theta)) \approx \theta + \rho + s(\theta) + \eta(\theta)$.

Adding a constant to the solution of 18.5 yields another solution. We can add such a constant to set $s(0) = 0$.

As they are periodic functions we can expand them in a Fourier series:

$$\eta(\theta) = \sum_{n \neq 0} \eta_n e^{in\theta}, \quad s(\theta) = \sum_n s_n e^{in\theta}$$

The $n = 0$ is absent in η because it has zero average.

$$[e^{in\rho} - 1]s_n = \eta_n, \quad n \neq 0 \tag{18.6}$$

We can determine s_0 by the condition that $s(0) = 0$.

$$s(\theta) = \sum_{n \neq 0} \eta_n \frac{e^{in\theta} - 1}{e^{in\rho} - 1}$$

If $\rho = \frac{m}{n} 2\pi$ for pair of integers, $e^{in\rho} - 1$ would vanish. Thus, to solve equation 18.5, $\frac{\rho}{2\pi}$ should not be a rational number. This is analogous to the condition in the last section that the diagonal entries of L should be unequal. Now we see how to define a measure of how far away w is from a rotation

$$\sup_\theta |s(\theta)| \leq \delta(\rho, \eta) \equiv \sum_{n \neq 0} \frac{2|\eta_n|}{|e^{in\rho} - 1|}$$

Given ρ, η_n, we thus set

$$w(\theta) = \theta + \rho + \sum_{n \neq 0} \eta_n e^{in\theta}, \quad S(\theta) = \theta + \sum_{n \neq 0} \eta_n \frac{e^{in\theta} - 1}{e^{in\rho} - 1}$$

Then we find $\tilde{w} = S^{-1} \circ w \circ S$ as the solution of

$$w(S(\theta)) = S(\tilde{w}(\theta))$$

This allows us to determine the new $\tilde{\rho}, \tilde{\eta}_n$ as

$$\tilde{\rho} = \frac{1}{2\pi} \int_0^{2\pi} [\tilde{w}(\theta) - \theta] d\theta, \quad \tilde{\eta}_n = \frac{1}{2\pi} \int_0^{2\pi} [\tilde{w}(\theta) - \theta] e^{in\theta} d\theta, \quad n \neq 0$$

If $\delta(\rho, \eta)$ is small enough, $\delta(\tilde{\rho}, \tilde{\eta})$ will be of order $\delta^2(\rho, \eta)$. By iterating this, we will converge to a pure rotation: the convergence is quadratic as is typical with Newton's method.

Example 18.5: The diffeomorphism

$$w(\theta) = \theta + \sqrt{2}\pi + \frac{1}{3}[\cos\theta - \sin(2\theta)]$$

has $\delta \approx 0.76$. It can be reduced to a rotation by $\lambda \approx 4.45272$ (to an accuracy $\delta < 10^{-10}$) in about seven iterations. The resulting rotation number λ is close to the initial value $\sqrt{2}\pi \approx 4.44288$.

So, we see that there is a neighborhood of an irrational rotation that is not chaotic. How big is this neighborhood for a given irrational number $\frac{\rho}{2\pi}$? This is a very interesting problem in number theory.

18.4. Diophantine approximation

Set $\rho = 2\pi r$. For any pair of integers m, n,

$$e^{2\pi i n r} = e^{2\pi i [nr - m]}$$

Thus

$$\left| e^{i n \rho} - 1 \right| = 2 |\sin \pi (nr - m)|$$

So the question is, how small can $nr - m$ get, for a given r? If r is rational, it can vanish. For irrational numbers, it can get quite small, when $r \approx \frac{m}{n}$.

An irrational number is said to be of Diophantine exponent a if

$$\left| r - \frac{m}{n} \right| > \frac{C(r, \epsilon)}{n^{a+\epsilon}}, \forall m, n \in \mathbb{Z}, \quad \epsilon > 0$$

for some C that can depend on r, ϵ but not m, n. Note the direction of the inequality. We are saying that the error in approximating r by rationals should not be too small.

There are some numbers (e.g., e or $\sum_{n=0}^{\infty} 10^{-n!}$) which do not satisfy this condition for any a, no matter how big. They are called Liouville numbers. They are not "irrational enough": they are too well-approximated by rationals. Such numbers are always either rational or transcendental (do not satisfy finite order polynomial equations with integer coefficients), but most transcendental numbers are believed to have a finite Diophantine exponent.

A theorem of Roth says that irrational algebraic numbers (solutions of polynomial equations with integer coefficients, like $\sqrt{2}$ or $\frac{1+\sqrt{5}}{2}$) are of exponent 2: the errors in approximating them by rationals only decrease as some power of the denominator. They are "irrational enough".

Note that $|\sin \pi x| > |2x|$ for $|x| < \frac{1}{2}$. Suppose r is type of (C, a). We can pick an m that makes $|nr - m|$ small for each n (certainly less than a half) to get

$$\left| e^{2\pi i n \rho} - 1 \right| = 2 |\sin[\pi(nr - m)]| > 4|nr - m| > \frac{4C}{n^{a-1}}$$

Thus

$$\delta(\rho, \eta_n) < \frac{1}{4C} \sum_{n \neq 0} |\eta_n| n^{a-1}$$

In order for $\delta(w)$ to be small, the high frequency components of w must fall off fast. The better the number $\frac{\rho}{2\pi}$ is approximable rationally (a large, C small), the more stringent the condition on the perturbation η. Thus, the rotations by such angles are more unstable: smaller corrections can become non-normalizable and hence chaotic.

The conclusion is that even small perturbations to a rotation $\theta \mapsto \theta + \rho$ can be chaotic if $\frac{\rho}{2\pi}$ is rational or is a Liouville number. But if $\frac{\rho}{2\pi}$ is an irrational algebraic number, there is a neighborhood of perturbations that remains non-chaotic.

18.5. Hamiltonian systems

Recall that an integrable system is one for which there is a canonical transformation $(p_i(P,Q), q^i(P,Q))$ such that the hamiltonian $H(p(P,Q), q(P,Q))$ depends only on the momentum (also called action) variables P_i. Such a canonical transformation is generated by $S(P,q)$ which solves the Hamilton–Jacobi equation 10.3. The functions $(p_i(P,Q), q^i(P,Q))$ are found by solving 10.2. The Hamilton's equations become 10.1. The frequencies $\omega^i(P)$ will depend on the quantities P. It is only for linear systems (harmonic oscillator) that they are constants in phase space.

If the orbit is bounded, these variables Q^i are angles. By convention they are chosen to have the range $[0, 2\pi]$. Thus, for each value of P_i there is an invariant torus whose coordinates are Q^i. The nature of the orbit on the invariant torus depends on whether the frequencies are commensurate. The frequency vector $\omega^i(P)$ is commensurate if there are integers such that

$$\sum_i n_i \omega^i(P) = 0$$

This means that some of the frequencies are in resonance. We will see that this case is unstable: there are small perturbations to the hamiltonian that can spoil the integrability. It is important that the frequencies depend on the action variables. So immediately close to resonant tori there will be others which are not resonant: changing I^i can lead to some small shift in frequency that takes us off resonance.

Example 18.6: For a system with just two degrees of freedom

$$n_1 \omega^1(I) + n_2 \omega^2(I) = 0$$

has a solution precisely when $\frac{\omega_2}{\omega_1}(I)$ is a rational number. In this case, the orbits are periodic. If there are no such integers, the orbit is dense: it can come as close to any point in the torus as you want.

18.5.1   Perturbations of integrable systems

Consider now a system that is close to being integrable. That is, we can find canonical variables such that the dependence on the angle variables is small. We can expand in a Fourier series

$$H(P,Q) = H_0(P) + \sum_{n \neq 0} H_n(P) e^{in \cdot Q} \equiv H_0(P) + h(P,Q)$$

As in the previous examples, even small perturbations can lead to chaos if there are resonances. The correct measure of the size of the perturbations turns out to be

$$\delta(H, P) = \sum_{n \neq 0} \frac{|H_n|}{|n \cdot \omega(P)|}$$

Note that this quantity depends not just on the hamiltonian H but also on the value of the nearly conserved quantities P. It is possible for the perturbation to have a small effect in some regions of the phase and a big (chaotic) effect in other regions. In the regions where $\delta(H, P)$ is small enough, we will be able to solve the Hamilton–Jacobi equation by Newton's iteration. Thus in those regions we would have found a canonical transformation that brings it to the integrable form and the system is not chaotic. The tricky part is that the regions where $\delta(H, P)$ is small and those where it is large are not well-separated: they interlace each other, so that the chaotic boundary is often some fractal.

So we seek a canonical transformation $(P.Q) \mapsto (P', Q')$ generated by some function $S(P', Q)$ satisfying the H–J equation.

$$H\left(\frac{\partial S(P', Q)}{\partial Q}, Q\right) = E$$

We solve this by Newton's iteration. As always we start with the solution of the linear approximation. We make the ansatz

$$S(P', Q) = P'_i Q^i + s(P', Q)$$

The first term simply generates the identity map. The second term is the correction.

$$H_0\left(P' + \frac{\partial s(P', Q)}{\partial Q}\right) + h\left(P' + \frac{\partial s(P', Q)}{\partial Q}, \phi\right) = E$$

The linear approximation is then

$$\omega^i(P')\frac{\partial s}{\partial Q^i} + h(P', Q) = \text{constant}$$

If we expand in Fourier series

$$s_n(P', Q) = \sum_{n \neq 0} s_n(P')e^{in.Q}$$

this becomes

$$in \cdot \omega(P')\, s_n(P') + h_n(P') = 0, \quad n \neq 0$$

The solution is

$$s_n(P') = i\frac{h_n(P')}{n \cdot \omega(P')}$$

Now we see that there will be a divergence due to resonance if the denominator becomes small. Also the quantity $\delta(H, P')$ that measures the size of the perturbation is simply the l^1 norm of the sequence s_n.

$$\sum_{n \neq 0} |s_n(P')| = \delta(H, P')$$

Now we make the change of variables implied by this generated function:

$$Q^{k\prime} = Q^k + \frac{\partial s(P',Q)}{\partial P^{k\prime}}, \quad P_k = P'_k + \frac{\partial s(P',Q)}{\partial Q^k}$$

We solve this equation (without any further approximations) to determine (P,Q) as functions of (P',Q'). substituting into H will give us a new hamiltonian

$$H'(P',Q') = H(P(P',Q'),Q(P',Q'))$$

If $\delta(H,P)$ is small enough, $\delta(H',P')$ will be even smaller. Typically it will be of order $\delta^2(H,P)$. So we can repeat the whole process over again, converging to some integrable hamiltonian, valid for the region where $\delta(H,P)$ is small.

There have been many refinements to this theory in recent years. It is possible to solve the problem by a convergent power series rather than an iterative procedure. Chaos can be shown to be limited in realistic problems in astronomy (Celletti and Chierchia, 2007) such as the Sun–Jupiter–Asteroid three body problem.

> **Problem 18.1:** Write a program that can find a root of a function of one variable using Newton's method. Try out the method for some choices of the function and the initial value.
>
> $$f(x) = x - 1.5\sin(x),$$
> $$f(x) = (1.166 + 1.574i) + (0.207 + 0.9726i)x + (0.7209 + 0.0099i)x^6$$

> **Problem 18.2:** Write a numerical program that diagonalizes a matrix with unequal diagonal entries using the method of this chapter. Test it on the example given above.

> **Problem 18.3:** Write a symbolic/numerical program that brings a diffeomorphism of the circle with small δ to the normal form. (The solution for $\tilde{w}$, as well as the evaluation of integrals, should be done numerically.) Test it on the example given above.

> **Problem 18.4:** *Research project**** Consider the circular restricted three body problem. Treat the smaller primary as causing the perturbation. Express the hamiltonian in terms of the action-angle variables of the two body problem (Delaunay). Implement the KAM iteration in a symbolic/numerical program. Show that the procedure converges to sufficient accuracy for a realistic problem (Celletti and Chierchia, 2007) (Sun–Jupiter–Victoria).

Further reading

Arnold, V.I. (1961). Small denominators: Mappings of the circumference onto itself. *AMS Translations*, **46(1965)**, 213–88.

Arnold, V.I. (1978). *Ordinary Differential Equations* (1st edn). MIT Press, Cambridge, MA.

Arnold, V.I. (1989). *Mathematical Methods of Classical Mechanics* (2nd edn). Springer, New York, NY.

Birkhoff, G.D. (1922). Surface transformations and their dynamical applications. *Acta Math.*, **43**, 1–119.

Celletti, A. and Chierchia, L. (2007). KAM stability and celestial mechanics. *Memoirs of the AMS*, **187(878)**, 1–134.

Connes, A. (1994). *Non-Commutative Geometry* (1st edn). Academic Press, San Diego, CA.

Feigenbaum, M. (1978). Quantitative universality for a class of non-linear transformations. *J. Stat. Phys.*, **19**, 25–52.

Feynman, R.P. (1948). Space-time approach to non-relativistic quantum mechanics. *Reviews of Modern Physics*, **20**, 367–87.

Landau, L.D. and Lifshitz, E.M. (1977). *Mechanics* (3rd edn). Pergamon Press, Oxford.

Lefschetz, N. (1977). *Differential Equations: Geometric Theory* (2nd edn). Dover Press, New York, NY.

McKean, H. and Moll, V. (1999). *Elliptic Curves: Function Theory, Geometry, Arithmetic* (1st edn). Cambridge University Press, Cambridge, UK.

McMullen, C.T. (1994). *Complex Dynamics and Renormalization* (1st edn). Princeton University Press, Princeton, NJ.

Milnor, J. (1999). *Dynamics in One Complex Variable* (1st edn). Vieweg, Wiesbaden.

Montgomery, R. (2005). Hyperbolic pants fit a three-body problem. *Ergodic Theory and Dynamical Systems*, **25**, 921–47.

Rajeev, S.G. (2008). Exact solution of the Landau–Lifshitz equations for a radiating charged particle in the Coulomb potential. *Ann. Phys.*, **323**, 2654–2661.

Strogarz, S.H. (2001). *Nonlinear Dynamics and Chaos* (1st edn). Westview Press, Boulder, CO.

Sudarshan, E.C.G. and Mukunda, N. (1974). *Classical Dynamics: A Modern Perspective* (1st edn). John Wiley, New York, USA.

Wayne, C.E. (2008). *An Introduction to KAM Theory*. Unpublished. Avaliable at http://math.bu.edu/people/cew/preprints/introkam.pdf.

Yoshida, H. (1990). Construction of higher order symplectic integrators. *Phys. Lett.*, **A150**, 262–8.

Index